Defense Modeling, Simulation, and Analysis

MEETING THE CHALLENGE

Committee on Modeling and Simulation for Defense Transformation

Board on Mathematical Sciences and Their Applications

Division on Engineering and Physical Sciences

NATIONAL RESEARCH COUNCIL
OF THE NATIONAL ACADEMIES

THE NATIONAL ACADEMIES PRESS
Washington, D.C.
www.nap.edu

THE NATIONAL ACADEMIES PRESS 500 Fifth Street, N.W. Washington, DC 20001

NOTICE: The project that is the subject of this report was approved by the Governing Board of the National Research Council, whose members are drawn from the councils of the National Academy of Sciences, the National Academy of Engineering, and the Institute of Medicine. The members of the committee responsible for the report were chosen for their special competences and with regard for appropriate balance.

This study was supported by Subcontract No. 172238J between the National Academy of Sciences and ITT Industries, AES Division under a prime contract with the Department of Navy. Any opinions, findings, conclusions, or recommendations expressed in this publication are those of the author(s) and do not necessarily reflect the views of the organizations or agencies that provided support for the project.

International Standard Book Number-10 0-309-10303-7
International Standard Book Number-13 978-0-309-10303-9

Additional copies of this report are available from the National Academies Press, 500 Fifth Street, N.W., Lockbox 285, Washington, DC 20055; (800) 624-6242 or (202) 334-3313 (in the Washington metropolitan area); Internet, http://www.nap.edu.

Printed in the United States of America

THE NATIONAL ACADEMIES

Advisers to the Nation on Science, Engineering, and Medicine

The **National Academy of Sciences** is a private, nonprofit, self-perpetuating society of distinguished scholars engaged in scientific and engineering research, dedicated to the furtherance of science and technology and to their use for the general welfare. Upon the authority of the charter granted to it by the Congress in 1863, the Academy has a mandate that requires it to advise the federal government on scientific and technical matters. Dr. Ralph J. Cicerone is president of the National Academy of Sciences.

The **National Academy of Engineering** was established in 1964, under the charter of the National Academy of Sciences, as a parallel organization of outstanding engineers. It is autonomous in its administration and in the selection of its members, sharing with the National Academy of Sciences the responsibility for advising the federal government. The National Academy of Engineering also sponsors engineering programs aimed at meeting national needs, encourages education and research, and recognizes the superior achievements of engineers. Dr. Wm. A. Wulf is president of the National Academy of Engineering.

The **Institute of Medicine** was established in 1970 by the National Academy of Sciences to secure the services of eminent members of appropriate professions in the examination of policy matters pertaining to the health of the public. The Institute acts under the responsibility given to the National Academy of Sciences by its congressional charter to be an adviser to the federal government and, upon its own initiative, to identify issues of medical care, research, and education. Dr. Harvey V. Fineberg is president of the Institute of Medicine.

The **National Research Council** was organized by the National Academy of Sciences in 1916 to associate the broad community of science and technology with the Academy's purposes of furthering knowledge and advising the federal government. Functioning in accordance with general policies determined by the Academy, the Council has become the principal operating agency of both the National Academy of Sciences and the National Academy of Engineering in providing services to the government, the public, and the scientific and engineering communities. The Council is administered jointly by both Academies and the Institute of Medicine. Dr. Ralph J. Cicerone and Dr. Wm. A. Wulf are chair and vice chair, respectively, of the National Research Council.

www.national-academies.org

COMMITTEE ON MODELING AND SIMULATION FOR DEFENSE TRANSFORMATION

SALLIE KELLER-McNULTY, Rice University, *Chair*
KIRSTIE L. BELLMAN, The Aerospace Corporation
KATHLEEN M. CARLEY, Carnegie Mellon University
PAUL K. DAVIS, RAND
RICHARD IVANETICH, Institute for Defense Analyses
KATHRYN B. LASKEY, George Mason University
R. BOWEN LOFTIN, Texas A&M University at Galveston
GEN DAVID M. MADDOX, Consultant
DENNIS K. McBRIDE, Potomac Institute for Policy Studies
COL MICHAEL McGINNIS, United States Military Academy
STEPHEN POLLOCK, University of Michigan
DAVID R. PRATT, SAIC
STEPHEN M. ROBINSON, University of Wisconsin-Madison
DETLOF von WINTERFELDT, University of Southern California
MICHAEL ZYDA, University of Southern California

Staff

SCOTT WEIDMAN, Director, Board on Mathematical Sciences and Their Applications
NEAL GLASSMAN, Senior Staff Officer
BARBARA WRIGHT, Administrative Assistant

For more information on BMSA, see its Web site at http://www7.nationalacademies.org/bms/, write to BMSA, National Research Council, 500 Fifth Street, N.W., Washington, DC 20001, call (202) 334-2421, or send e-mail to bms@nas.edu.

Acknowledgments

This report has been reviewed in draft form by individuals chosen for their diverse perspectives and technical expertise, in accordance with procedures approved by the National Research Council's Report Review Committee. The purpose of this independent review is to provide candid and critical comments that will assist the institution in making its published report as sound as possible and to ensure that the report meets institutional standards for objectivity, evidence, and responsiveness to the study charge. The review comments and draft manuscript remain confidential to protect the integrity of the deliberative process. We wish to thank the following individuals for their review of this report:

Richard Deckro, Air Force Institute of Technology,
Paul Fishwick, University of Florida,
Robert A. Frosch, Harvard University,
Thom J. Hodgson, North Carolina State University,
Gregory S. Parnell, United States Military Academy,
Ervin Y. Rodin, Washington University, and
Marlin Thomas, Air Force Institute of Technology.

Although the reviewers listed above have provided many constructive comments and suggestions, they were not asked to endorse the conclusions or recommendations, nor did they see the final draft of the report before its release. The review of this report was overseen by John Christie of the Logistics Management Institute. Appointed by the National Research Council, he was responsible for making certain that an independent examination of this report was carried out in accordance with institutional procedures and that all review comments were carefully considered. Responsibility for the final content of this report rests entirely with the authoring committee and the institution.

The committee acknowledges the valuable contribution of the following individuals who provided input at the committee's meetings:

John Allen, Defense Advanced Research Projects Agency,
Michael Bailey, U.S. Marine Corps, Training and Education Command,
Philip Barry, George Mason University,
Donald R. Bates, Office of the Secretary of Defense, Joint Warfare Systems Office,
ADM Arthur Cebrowski, DoD Office of Force Transformation,
Anthony Cerri, U.S. Joint Forces Command,
David Chu, Under Secretary of Defense for Personnel and Readiness,
Philip Coyle, former head of DoD Office of Test and Evaluation,

W. Forrest Crain, Headquarters, Department of the Army,
David Davis, George Mason University,
Fred Hartman, Office of the Under Secretary of Defense for Personnel and Readiness,
Orville Hay, U.S. Joint Forces Command,
Jacqueline Henningsen, Air Force Studies and Analysis Agency,
Randy Hill, Institute for Creative Technologies, University of Southern California,
Andrew Loerch, Mitre Corporation,
Gregory Melcher, Chief of Naval Operations Staff,
Randy Michelsen, Los Alamos National Laboratory,
Steven Moore, U.S. Joint Forces Command,
Al Sciarretta, CNS Technologies, Inc.,
John Surdu, Analysis Center of the U.S. Army Training and Doctrine Command, and
David Weddel, Space and Naval Warfare Systems Command.

The committee also gained much by sending individual members on site visits to Purdue University, Los Alamos National Laboratory, the Sante Fe Institute, and the NATO Modeling Center in Hampton, Virginia, and is very appreciative of the exchanges of expertise and hospitality shown at these visits.

The committee also thanks the following experts who contributed to a session in October 2004 to better scope the study:

David Banks, Duke University,
Thomas Cioppa, Naval Postgraduate School,
John Sommerer, Applied Physics Laboratory, Johns Hopkins University,
Stuart Starr, Barcroft Research Institute, and
Marlin Thomas, Purdue University.

Contents

Executive Summary

Modeling, simulation, and analysis (MS&A) is a crucial tool for military affairs. While DoD's reliance on MS&A, in some form, dates back to World War I, changes in the enterprise of MS&A have not kept pace with the new demands arising from rapid changes in DoD processes and missions or with the rapid changes in the technology available to meet those demands. This report, commissioned by the Defense Modeling and Simulation Office[1] and written by the Committee on Modeling and Simulation for Defense Transformation, identifies these shortcomings and suggests where and how they should be addressed. Although many other studies speak of modeling and simulation as a unit, the development of models without regard for planned subsequent analysis is a flawed process. Consequently, there is a need for examination and recommendations concerning all three: modeling, simulation, and analysis.

The charge to the committee, given in Chapter 1, has four tasks:

- Evaluate the current capabilities of MS&A to support defense transformation.
- Identify high-leverage opportunities for modeling and simulation (M&S) research.
- Recommend approaches for framing existing and future MS&A products to decision makers and giving them the appropriate context and caveats, including evaluation of the risks.
- Identify potential organizational and human resource development issues related to these themes.

The committee regards these tasks as a charge to recommend steps DoD must take to improve its entire MS&A infrastructure: the science base of MS&A; the execution of MS&A, including the interface of its practitioners with decision makers who may not be experts in the field; and the education and training of DoD MS&A practitioners. These are the subjects of Chapters 3, 4, and 5. The recommendations in this report are of two types: those pointing to promising areas for investigation and those pointing to procedural changes or suggestions for professional practice. The first type of recommendation involves research, where progress is unpredictable and, to some extent, a function of the resources applied. The second type requires commitment of persons or organizations in authority in DoD, and suggested commitments are noted in the text. The committee believes a holistic approach is needed to improve the MS&A enterprise, addressing all of these aspects of infrastructure simultaneously. In Chapter 6, it identifies the M&S communities at DoD that should be concerned with its separate recommendations.

Chapter 2 examines the changing mission of DoD and the changing environment in which it has been called on to operate. In addition to preparing for traditional warfare, DoD also must prepare for irregular conflicts, police actions, stabilization and reconstruction missions, and other nontraditional activities. This preparation must be done in a strategic environment that lacks the clarity of the cold war. Decision making at DoD increasingly considers the entire range of diplomatic, intelligence, and economic options, as well as military force—the so-called "DIME space." At the same time, new technology has emerged. For instance, the autonomous operation of separate military units has been replaced by military systems that are increasingly interconnected and interdependent. The growing prevalence of embedded software, which often includes models and simulations, presents new challenges. Networks are now ubiquitous, and their use has become both an essential tool in the development of military strategy and an essential element of the strategy itself. Understanding the behavior of networks as they grow more complex is a difficult scientific and technical challenge. The combined operation of unmanned and manned systems may give rise to unforeseen and unmodeled emergent behavior. Many key aspects of this changing military landscape are

[1]Recently renamed the M&S Coordination Office.

not well understood, and existing knowledge about phenomena such as terrorism or counterinsurgency has not been fully codified into models. Accordingly, models and analysis of networks and of complex adaptive systems, along with the use of embedded M&S, are central drivers for the development of future MS&A tools. The way that the defense MS&A enterprise should respond to these developments—the emergence of networks, adaptive systems, and embedded systems—forms the central theme of this report.

Chapter 3 identifies high-leverage opportunities for MS&A research needed to address the expanded mission space. Such research would expand the science foundation of MS&A. Building on a list of the new capabilities needed for the MS&A enterprise, the chapter discusses promising approaches for obtaining them, emphasizing the mathematical, scientific, and computational advances that have not yet been sufficiently incorporated into modeling technology and new topics where research can have a disproportionately large payoff.

The committee concluded that four main objectives should guide DoD's MS&A efforts, and it developed the following four general recommendations:

Recommendation 1: DoD should give priority to developing flexible, adaptive, and robust MS&A methods for evaluating military strategies.

Recommendation 2: DoD should ensure that the basic architecture of its MS&A systems reflects modern concepts of network-centric warfare.

Recommendation 3: DoD should give special emphasis to the development of MS&A capabilities that are needed within embedded systems.

Recommendation 4: DoD should establish a comprehensive and systematic approach for developing the MS&A capabilities to represent network-centric operations:

- **Enhance and sustain collaborations among the various parties developing network-centric MS&A capabilities.**
- **Continue and extend the development of existing approaches to modeling network-centric operation.**
- **Establish a new mathematical basis for models describing network-centric operation, drawing on an array of approaches, particularly complex, adaptive systems research.**

The diversity of challenges facing the DoD MS&A community requires a diversity of mathematical and modeling approaches, and the committee recommends that this be embraced as a guiding principle.

Recommendation 5: DoD's analytical organizations should take a portfolio approach to designing their analyses and supporting research, investing in a range of methods including diverse models, games, field experiments, and other ways to obtain information.

In Chapter 3, the committee goes on to develop four more recommendations related to particular technical directions identified in the chapter:

Recommendation 6: DoD should devote significant research to social behavioral networks and multiagent systems because both are promising approaches to the difficult modeling challenges it faces.

Recommendation 7: DoD should form a research center or consortium focused on game-based training and simulation.

This report highlights the ubiquity of network-centricity for future DoD operations, but there is little scientific foundation for DoD networks or for those that appear elsewhere throughout our society. The committee believes that such foundational research would redound to the benefit of DoD.

Recommendation 8: DoD should support and extend initiatives to cooperate with other agencies funding research on networks.

Recommendation 9: DoD should begin cooperative programs of research into embedded systems with other agencies facing similar demands.

In Chapter 4 the committee presents approaches for improving the interface between MS&A practitioners and decision makers and makes two related recommendations. This interface presents a perennial challenge, not just in DoD, but in every situation where the results of sophisticated and nuanced MS&A must be condensed and conveyed, at appropriate but variable levels of detail, to end users. As DoD plans for its expanded mission space in the face of increased uncertainties, it is critical that the best MS&A be used and used effectively, because novel missions and increased uncertainties lessen the ability of decision makers to rely solely on their collected experience and judgment. Good professional practices are identified for translating a decision problem into an MS&A study that will assist in making that decision and for facilitating appropriate interaction with the decision maker. Emphasis is put on the communication skills needed to frame results accurately, including representation of the many levels of uncertainty inherent in any such solution.

There is a fundamental need for better understanding of the cognitive styles of decision makers and their interaction with different forms of MS&A. For instance, much is unknown about how information is absorbed and what biases might be introduced by alternative means of presentation. Research into the different cognitive styles of decision makers could help the MS&A community understand these styles and improve the practice of decision making by using MS&A. The committee has developed the following two recommendations to address this concern:

Recommendation 10: DoD should strive to better understand the cognitive styles of decision makers and their interaction with different forms of MS&A. Research into decision-making styles would improve decision making with MS&A by affording intuition and insight into complex problems and enhancing the creativity employed in their solution.

DoD's expanded mission space includes more and greater uncertainties than conventional warfare. Because of the greater underlying uncertainty of the phenomena we now need to model and analyze, it is increasingly important to aim for MS&A methods that are able to cope with that uncertainty:

Recommendation 11: DoD should seek better methods to characterize, quantify, and manage the uncertainty inherent in all aspects of MS&A—including inputs, modeling assumptions, parameters, and options.

Chapter 5 discusses the training and continuing education of MS&A practitioners. The new defense challenges call for more widespread understanding of new MS&A methods and, in general, a high standard of professionalism across a broader swath of MS&A people. This high level of professionalism will give decision makers more confidence in the ability of MS&A people to contribute meaningfully to very difficult decisions that should not be based on incomplete experience. A core curriculum is proposed, the variety of skills needed by an MS&A team are identified, and ways to ensure that both military and civilian MS&A practitioners remain current are suggested. In addition, the committee points out that the talents required of MS&A practitioners and the roles they play in the process may change over the course of an MS&A study. These discussions led to two recommendations:

Recommendation 12: DoD must give its MS&A practitioners some exposure to all the topics in the core curriculum: the MS&A life cycle; continuous and discrete simulation; probability and statistics; topics in computing; deterministic modeling and optimization; MS&A evaluation; human-simulation interaction; modeling humans; and managing MS&A. Familiarity with these topics is essential for all practitioners, although the depth of knowledge needed will depend on a practitioner's particular role.

Recommendation 13: DoD should ensure that its educational programs provide experiences in which students integrate the activities of modeling, simulation, analysis, evaluation, and communication to address real-world problems of importance to the consumers of the information.

Finally, in Chapter 6, the committee examines the need for coordinated research in the base of military science to support MS&A:

Recommendation 14: DoD should identify (or create) and charge an organization with responsibility for developing and supporting a program of research and development directed at improving and updating the base of military science for combat and noncombat modeling. That same organization would be responsible for effecting the recommendations on education that are called for in Chapter 5.

The committee discussed the need for research in numerous promising areas but has chosen to recommend only five: social behavioral networks, game-based training and simulation, cognitive decision making, network science, and embedded systems. It further recommends that the last two areas be supported jointly with other agencies that have a common interest in them. The committee regards these five as the most important in the presence of constrained resources. Having recognized the need for an office to coordinate improvements to DoD MS&A, the committee concludes Chapter 6 by suggesting which DoD MS&A communities should be responsible for implementing its recommendations.

1

Introduction

MOTIVATION FOR THE REPORT

The evolving missions of the Department of Defense (DoD) put heavy and new demands on DoD's modeling, simulation, and analysis (MS&A) capabilities. Some of the most noteworthy facets of this evolution are these:

- *Operations in a broader context.* It is increasingly common for DoD planners to consider the entire range of diplomatic, intelligence, military, and economic (DIME) means for achieving national goals so as to enable what is called effects-based planning and effects-based operations (EBO).[1] Thus, there is a need for MS&A capabilities that can address this broader context in a coordinated and sensible manner while taking into account the learning and adaptation that take place en route to attaining the desired effects.
- *Capabilities-based planning.* DoD has shifted to capabilities-based planning (CBP), which stresses the development and honing of capabilities that can be applied to a wide range of challenges and circumstances (Rumsfeld, 2006; Davis, 2002), and to CBP's analogue in operations, adaptive planning, in which war plans are developed so as to anticipate and then support major changes as circumstances change.
- *Increased networking.* The increasing interconnectedness of forces, often referred to as network-centric operations, offers many advantages but also presents planning and operational challenges.[2] For a more complete discussion of network-centric operations, see the section "Representing Networking" in Chapter 3.
- *Involvement in nontraditional warfare.* As U.S. military prowess in traditional combat becomes more and more overwhelming, adversaries have been adopting strategies and tactics designed to avoid traditional engagements.[3,4] These include irregular warfare, such as characterizes the insurgency in Iraq; moving more installations underground or otherwise seeking to frustrate overhead surveillance and precision targeting; mingling with the civil population; and attacking the United States, its allies, and their interests at home.
- *Stabilization and reconstruction operations.* The U.S. military has not traditionally had much enthusiasm for stability and reconstruction missions, but these are often important and are likely to become an important element of military operations in the foreseeable future. Success in these missions without the benefit of an overwhelming force presence poses daunting challenges.

Planning for nontraditional missions and transformed forces requires analyses of the capabilities and trade-offs for various proposed force structures and equipment and also

[1]Some in the DoD use the acronym DIME as shorthand for the space of diplomatic, intelligence, military, and economic actions. A related acronym is PMESII (political, military, economic, social, infrastructure, and information), which refers to the range of effects that planning and operations are meant to produce.

[2]See, for example, Alberts and Hayes (2003); Alberts et al. (1999); NRC (2000).

[3]Although the term "nontraditional mission" is often used, the strategies and tactics involved are as old as war itself. DoD often refers to "asymmetric warfare."

[4]Some DoD policy documents draw distinctions between planning for traditional, irregular, disruptive, and catastrophic conflict. Irregular threats are those posed by circumstances such as terrorism, insurgencies, or coercion. Disruptive threats are those that employ new capabilities, such as cyberattacks, directed-energy weapons, or attacks on assets in space. Catastrophic threats use weapons of mass destruction outside the normal battlespace to mount a major attack on national symbols or on infrastructure. The taxonomy builds on work by the Office of Force Transformation, whose first director was the late Admiral Art Cebrowski, who briefed the panel in 2004. The relevant Web site is http://www.oft.osd.mil.

requires doctrine for a range of military operations, including those with which the United States has very little or no experience. For these, as well as for more familiar operations, analysis based on M&S can be an invaluable tool for developing and assessing concepts across a broad set of circumstances.

STUDY GOALS AND PROCESS

To develop an overall vision for MS&A that can contribute to the fulfillment of evolving DoD needs, the Defense Modeling and Simulation Office[5] asked the National Research Council (NRC) to establish a committee on modeling and simulation for defense transformation and gave it the following charge:

> The study will make recommendations to guide the development of defense MS&A to better support defense decision makers at all levels. Specifically, the study will:
> —Evaluate the current capabilities of MS&A to support defense transformation.
> —Identify high-leverage opportunities for M&S research.
> —Recommend approaches for framing existing and future MS&A products to decision-makers, giving them the appropriate context and caveats, including evaluation of the risks.
> —Identify potential organizational and human resource development issues related to these themes.

Defense MS&A covers a very broad spectrum of technologies, from rough models created with spreadsheet software by one person to guide his or her own decisions, to billion-dollar simulation tools based on multiyear efforts by large teams to support completely separate groups of analysts and decision makers. Models range from those supported by years of use and validation to those just emerging from ongoing research. The committee that was formed—the Committee on Modeling and Simulation for Defense Transformation—focuses on the latter: the frontier of MS&A capabilities. This frontier regime of MS&A can gain the most from an outside study such as this one while still addressing the communication, organizational, and human resource issues called out in the charge. The committee's emphasis is on challenges facing the developers of new MS&A capabilities rather than on issues that influence, say, the transition from R&D to commercial tools or the effective and appropriate use of commercial simulation products.

This report builds on a number of others—by the NRC, the Defense Science Board, and others—over the course of more than a decade. One such report (NRC, 2002) includes an appendix summarizing 10 reports, from 1994 to 2000, on just the topic of M&S applied to acquisition. One of the best statements of issues in M&S is contained in Volume 9 of another NRC report (NRC, 1997).

The current report builds on the 1997 report in the following ways: (1) by dealing with issues that affect all of the services as well as joint operations; (2) by including considerations of how M&S tools are used in analysis for, ultimately, informing strategic decision making; (3) by discussing how current MS&A efforts across DoD can be adjusted to address the broader range of threats now accepted as being in the military's purview; and (4) by evaluating some of the most important new conceptual and technological approaches in MS&A.

This report provides high-level guidance for developing the MS&A capabilities that will support planning by DoD and the services in an evolving environment.

THE SCIENTIFIC AND SOCIETAL ROLE OF MS&A

The National Science Foundation (NSF) named a panel on simulation-based engineering science that held two workshops, in April 2004 and September 2005, and in their course made recommendations to NSF. The recently published report from these two workshops (NSF, 2006) examines the role of MS&A throughout our society, with emphasis on engineering applied to medicine, homeland security, energy, environment, materials, industry, and, very briefly, defense. After explaining the importance of MS&A for present and future U.S. competitiveness, the report makes recommendations to NSF that would allow NSF to advance the science and practice of MS&A. Although the NSF panel and this committee operated independently, the similarity of their recommendations is striking. Within the wide purview of MS&A, the challenges and opportunities facing DoD are those facing U.S. society as a whole.

In a sense, the current study can be regarded as a focused effort to examine the problems and solutions as they relate to DoD and in more detail than was possible in the NSF study. In Chapter 6, the committee reviews the NSF study and discusses the mutually reinforcing sets of recommendations.

THE ROLE OF MS&A IN THE DEPARTMENT OF DEFENSE

DoD's reliance on MS&A is not new. Ever since World War I, military analysis in some form has played an essential role in defense, with many success stories in the three domains of acquisition, training and operations, and defense planning (the last is sometimes inappropriately referred to as the "analysis" domain). In acquisition, for example, exceedingly accurate and reliable ballistic missiles have been acquired and fielded while relying heavily upon M&S, in part because realistic and exhaustive field testing is infeasible and too expensive. In training, cockpit simulators have been successful for decades, allowing personnel to develop skills and intuition in a safe and controlled fashion. The dramatic performance of U.S. armored-force units in the 1991 war with Iraq demonstrated to participants the great value of

[5]Recently renamed the M&S Coordination Office (M&SCO).

simulation-based training. MS&A is being effectively used for, among other things, (1) scheduling maintenance, (2) placing sonobuoys, (3) loading cargo, (4) developing training schedules, (5) search and detection, and (6) planning for statistically significant operational tests. Today's large-scale joint exercises routinely mix live activities (pilots flying, tankers maneuvering, etc.) with simulated activities, often seamlessly and often with participants not even knowing or caring which targets are real and which are simulated. Detailed mission-rehearsal simulations are now an essential and proven part of operations planning generally, as are simulations of mobilization, deployment, maneuver, and logistics. Today's defense-planning analysis draws on sophisticated end-to-end simulations of how alternative joint forces would actually be employed in hypothetical scenarios.

Although it is beyond the scope of this report to provide a complete history of DoD M&S, the committee lists below some early examples of models used successfully by DoD, versions of which are still extant:

- *Joint Theater Level Simulation (JTLS).*[6,7] JTLS is in many ways the grandfather of all DoD combat models. Development was begun in 1983 as a means to automate the board-game-based wargaming program that was then used by the Army. It has grown over time to represent the spectrum of joint and coalition operations.
- *SUPPRESSOR.*[8] This Air Force model started in 1978 and has been continuously updated to reflect changes in tactics and equipment. It is currently one of the most widely used air combat mission models and is also used for design and acquisition studies.
- *Corps Battle Simulation.*[9] This model and its successors have been in place for over 20 years and have been extended to include maneuver, command and control, fire support, air defense, combat service support, mobility, countermobility, survivability, intelligence, special operations, and psychological operations. It is currently used as the land warfare component of various joint training exercises.

While many past studies examined defense M&S, this study examines military analysis as well, because all three activities—modeling, simulation, and analysis—are essential, intertwined inputs to decision making. Models (mathematical or otherwise logical representations of entities, relationships, or processes of importance to military operations) and simulations (exercises that include computations based on those models and possibly humans executing related tasks) are most effective when they are designed with analyses in mind. It is best for military analysts to provide input into the creation of the models and simulations that will underpin their analyses, and the best M&S-based support for decision makers comes about when M&S personnel and military analysts work coherently to explore and illuminate the issues facing those decision makers. Because of the need for such rich connections between M&S and military analysis, this report is written as though there were a single MS&A community, which would be a desirable situation.[10]

MS&A is of value in the early stages of defense modernization, when roughly defined concepts can be examined and adjusted in virtual worlds. It might be of even greater importance as planning becomes more concrete, because MS&A will then be needed for detailed engineering-based or physics-based analyses of proposed equipment, forces, doctrinal and strategic choices, with whatever analytical technology exists. Take, for example, network-enabled forces. All the services are looking to network their forces as a way of ensuring information superiority, improving situational awareness, shortening the response time for military actions, and increasing synchronization and related effectiveness. But tools do not yet exist to explore the trade-offs associated with actually building network-enabled forces, and no one's intuition is adequate for the task. Networks and networked forces, because of their complexity, cannot simply be assembled from existing tools and expected to work as predicted—there are far too many unforeseen interactions, vulnerabilities, and feedback loops. Therefore, it is critical to develop capabilities for modeling, simulating, and analyzing the behavior expected from various networks and networked forces, based on particular choices of components, protocols, doctrine, attacks, and contexts.

In tandem with advanced MS&A's use to engineer transformed forces, it is also essential to developing budgets, personnel plans, logistics support plans, and so on for how the forces will be fielded and used. That is, models are needed that can represent all aspects of military operations. M&S will also be of central importance for training personnel on new equipment for new scenarios. Moreover, novel force structures are also effecting dramatic change in the military at the social and cultural levels, and M&S can prepare the ground (consciously or not) for future changes. MS&A can also help assess, support, and identify how to constrain or enable this evolutionary process.

The DoD's *Transformation Planning Guidance* (DoD, 2003, p. 8) affirms the need for strengthening and expanding

[6]http://www.rolands.com/Home/Projects/Project_index.htm.

[7]http://www.jfcom.mil/about/fact_jtls.htm.

[8]https://www.afsaa.hq.af.mil/content/SUPPRESSOR_Overview_Document.pdf.

[9]http://www.peostri.army.mil/PRODUCTS/CBS/.

[10]M&S is not, strictly speaking, a separate discipline that creates stand-alone tools, although there are certainly specialists within M&S who delve deeply to create new capabilities that can be incorporated into valuable tools. Similarly, military analysts are not always able to use canned software or predefined models if they are to perform well-targeted analysis.

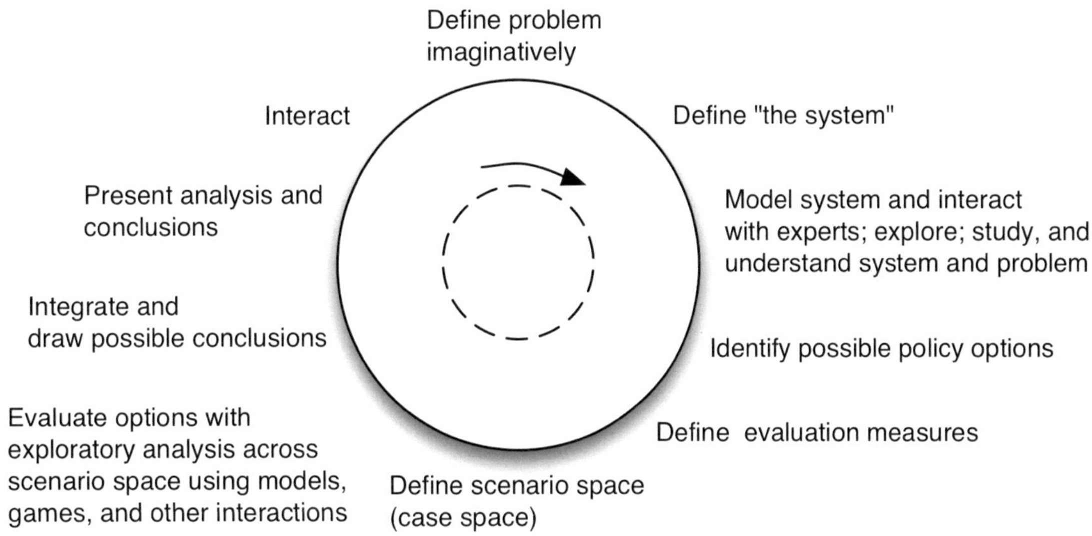

FIGURE 1.1 The MS&A modeling process.

MS&A capabilities. That document notes, in a paragraph on transformed strategic analysis, that

> . . . the Department needs a transformed analytic capability that can identify and assess risks for strategic planning. . . . DoD must be able to support a capabilities-based planning process that accounts for greater uncertainty in threats and capabilities, and must be capable of comparing risks across time and between multiple theater-level operations.

Strong MS&A capabilities are also needed to support the *Quadrennial Defense Review* (QDR) process, which periodically provides a detailed understanding of the posture of the military forces and how they are going to approach the changing world over the next 4 years. The main goals of the QDR are to understand four things: (1) the impacts of new technologies, (2) the impacts of the changing world on force structure and capabilities, (3) the changes in our adversaries, and (4) the changes in the budget.

While the use of MS&A is not explicitly called for in the QDR, it is clearly necessary if the objectives of the QDR are to be met. Unfortunately, the vast majority of the high-level, strategic models in common use within DoD are cold war legacies that do not reflect the full spectrum of current military missions and threats or the impact of civilians and civilian organizations.

ROLES PLAYED BY INDIVIDUALS IN THE MS&A ENTERPRISE

It is useful to distinguish various roles that are relevant to the MS&A enterprise. The committee has identified five such roles. In practice, however, the functions of each role may be divided among several people or, conversely, one person may take several roles. The modeling process into which the identified roles fit is pictured in Figure 1.1.

The roles and their associated functions are as follows:

- *Analysts* create the formal model representation from the real-world problem, act as domain experts, and interpret and present results.
- *Modelers and programmers* translate the representation into a documented and executable form.
- Implementers develop and execute the experimental plan and transform the model's raw outputs into useable results.
- *Managers* oversee the MS&A team by managing personnel, making required purchases necessary for efficient operation of the team, checking for quality of the MS&A product, and interacting with others involved in M&S governance.
- *Consumers* employ MS&A to support military decisions.

The effectiveness of MS&A in supporting military decisions depends heavily on how well practitioners play their respective roles and also on how well individuals in different roles work together toward a common objective. In its interactions with the MS&A community, the committee observed that different roles sometimes do not mesh as well as they should and that the quality of decisions can suffer as a consequence. Well-designed education and training can contribute to a solution.

STRUCTURE OF THIS REPORT

The increasing prominence of MS&A calls for a complete rethinking of systems, organizations, policies, and training, in the same way that the pervasive use of information technology has created the need for a reengineering of business processes. MS&A is becoming embedded throughout the force structure for enabling better decisions and control, presumably because it provides more complete information and offers well-validated, easily interpreted multifaceted models. All decision making—from strategic, long-term decisions to real-time command and control—has the potential to be better informed and can therefore be improved through the effective employment of MS&A.

Chapter 2 goes into greater detail about changes in DoD's operations and mission, how these changes place very challenging demands on MS&A, and how emerging IT-based technologies add other demands. Chapter 3 explores some of the ways in which R&D and technology have begun addressing these challenges. It discusses some specific directions that promise to provide the needed functionality and identifies some R&D topics that need further exploration in order to bring that promise to fruition. The chapter also discusses infrastructure needs that must be addressed to enable these technical advances. Chapter 4 discusses how to improve the effectiveness of the support the MS&A community provides to decision makers, and Chapter 5 discusses the education, training, and career development needed by the MS&A community. The report concludes with a chapter explaining how these promising directions in MS&A can be of maximal value to DoD.

Appendixes A-C contain further details on subjects that the committee felt could not be dealt with adequately in the body of the report.

REFERENCES

Alberts, D.S., J.J. Gartska, and F.P. Stein. 1999. *Network Centric Warfare—Developing and Leveraging Information Superiority.* 2nd ed. Command and Control Research Program Publication Series. Washington, D.C.: Department of Defense.

Alberts, David S., and Richard E. Hayes. 2005. *Power to the Edge: Command and Control in the Information Age*. Springfield, Va.: EBR Inc. Available at http://www.dodccrp.org/publications/pdf/Alberts_Power.pdf.

Davis, Paul K. 2002. *Analytic Architecture for Capabilities-Based Planning, Mission-System Analysis, and Transformation.* Santa Monica, Calif.: RAND.

Department of Defense (DoD). 2003. *Transformation Planning Guidance.*

Gordon, Steven C. 2000. "Determining the value of simulation," in *Proceedings of the Summer Simulation Conference.* Available at http:www.scs.org/getDoc.dfm?id=1126.

National Research Council (NRC). 1997. *Technology for the United States Navy and Marine Corps, 2000-2035, Becoming a 21st-Century Force.* Vol. 9, *Modeling and Simulation.* Washington, D.C.: National Academy Press.

NRC. 2000. *Network-Centric Naval Forces: A Transition Strategy for Enhancing Operational Capabilities.* Washington, D.C.: National Academy Press.

NRC. 2002. *Modeling and Simulation in Manufacturing and Defense Acquisition: Pathways to Success.* Washington, D.C.: National Academy Press.

National Science Foundation (NSF). 2006. *Simulation-Based Engineering Science.* Available at www.nsf.gov/pubs/reports/sbes-final-report.pdf.

Rumsfeld, Donald. 2006. *Quadrennial Defense Review.* Department of Defense.

2

The Changing Landscape

The changing landscape and its associated needs for model-supported analysis, as described in this chapter and in Chapter 3, reveal three overarching themes:

- DoD needs MS&A appropriate to complex, dynamic, adaptive systems because such systems pervade military combat, other aspects of military operations, and other political, military, economic, social, infrastructure, and information (PMESII) phenomena of interest.
- DoD needs MS&A that is capable of effectively representing the ubiquity and significance of networking.
- DoD needs new methods to design and model real-time simulations coupled to embedded devices.

This does not mean that all DoD MS&A should focus on complex, dynamic, adaptive systems or networks but means rather that DoD should consider them when developing and exploiting M&S for analysis and operations. Such considerations will affect the portfolio of investments, the terms of reference for individual projects, and the way in which integrative activities are undertaken. DoD's future policies and practices in the development and use of MS&A should explicitly address the growing role and contributions of these three overarching themes.

The remainder of this chapter elaborates on the changes faced by DoD. In Chapter 3, the committee makes specific recommendations for future directions in military MS&A and identifies the main challenges for MS&A in the upcoming decades.

The modernization of DoD is a direct result of the changed environment in which it has been called on to operate since the end of the cold war. While military planners must continue to prepare for engagements with the armed forces of other nations, there is an increasing necessity to plan against insurgents and terrorists as well. The range of missions has expanded from force-on-force engagements to also include counterterrorism, stability and support operations, and humanitarian relief. This has had an impact on force structure, weapon systems, equipment, and personnel across all the services. The environments in which these missions are performed routinely involve urban areas in which large numbers of noncombatants are present. Clearly these missions cannot be successfully accomplished by DoD alone—they must be performed using all the elements of national power, including diplomatic, economic, social, information, and military power. These new realities have given rise to organizational changes that increase force deployability and change doctrine to include effects-based operations.

This chapter examines the changed environment by looking first at new challenges that MS&A must be able to address and then at the new technological landscape that MS&A must represent. It concludes with a discussion of what is known about the return on investment (ROI) of MS&A in a military context.

NEW CHALLENGES FOR RESOURCE ALLOCATION, PLANNING, TRAINING, AND OPERATIONS

DoD's *Transformation Planning Guidance* (DoD, 2003, pp. 8 and 19) poses new challenges for resource allocation, planning, training, and operations, while affirming the need for strengthening and expanding MS&A capabilities. It notes, in a paragraph on "transformed strategic analysis," that the Department needs a transformed analytic capability that can identify and assess risks for strategic planning:

> . . . DoD must be able to support a capabilities-based planning process that accounts for greater uncertainty in threats and capabilities, and must be capable of comparing risks across time and between multiple theater-level operations.

Later, in identifying M&S as one of the key elements of infrastructure for concept development, it notes that

> a new generation of M&S is needed to support concept development [including] linking together many types of simu-

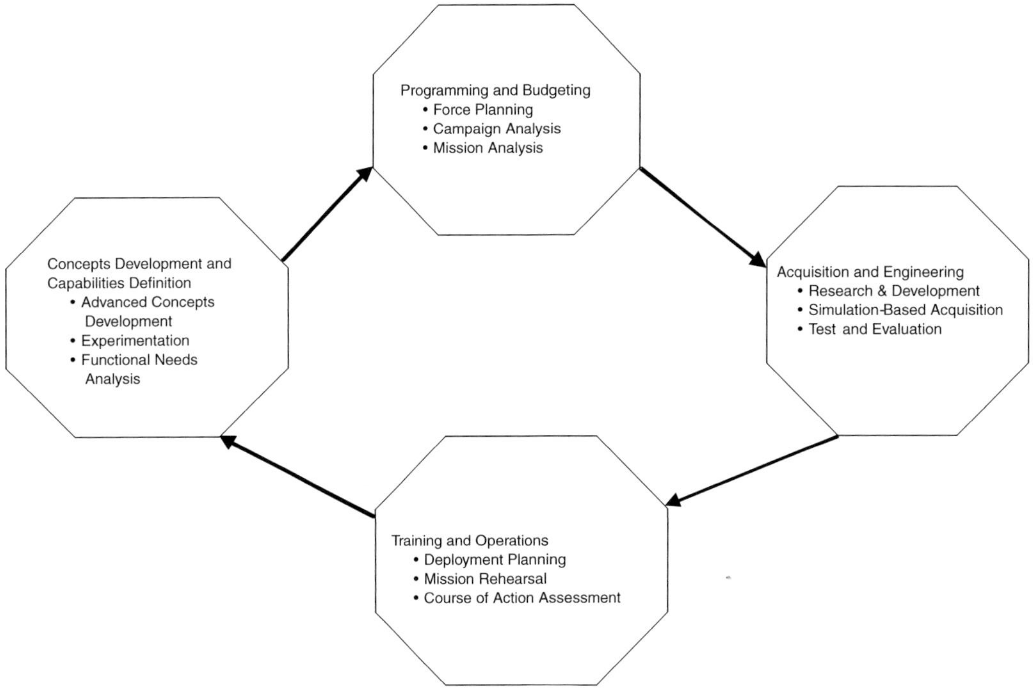

FIGURE 2.1 The interrelationship among DoD processes areas.

lations, from aggregate and detailed computer models to simulators and man-in-the-loop hardware components.

Integration of Department of Defense Processes

One of the challenges implied by that vision is the need for greater integration of DoD processes, which can be facilitated by improvements in MS&A. Three main DoD processes together lead to the development of military force capability:

- *Concept development and capabilities definition.* Currently called the Joint Capability Integration and Development System (JCIDS).
- *Programming and budgeting.* Currently called the Planning, Programming, Budgeting, and Execution process.
- *Acquisition and engineering.* Currently called the Defense Acquisition System and codified in DoDD 5000.1 and DoDI 5000.2.

The capability output from these processes is then employed in a fourth process, the actual training and operations of the forces. MS&A supports each of these four processes, as illustrated in Figure 2.1.

These processes cannot stand in isolation from one another but rather must be integrated to produce the overall capabilities. The cycle begins with concept development and capabilities definition specifying necessary force capabilities, which are supported through programming and budgeting, with the resources being used in acquisition and engineering to develop the materiel capabilities, which are then applied in the force through training and operations, with the output of this process then feeding back to a new cycle.

Ideally, however, the interaction should be more complex than is shown in Figure 2.1, with numerous feedback loops. To cite one particularly important example, there must be continual iteration among capabilities definition, programming and budgeting, and acquisition to allocate resources. That is, capabilities definition must use the outputs of *acquisition* to incorporate the cost of developing a particular capability, which is then weighed against other costs in programming and budgeting to estimate the resources that can be allocated to the particular capability. If the full resources cannot be allocated, then capabilities definition and acquisition must interact further to examine trade-offs and determine a fallback position. This, in turn, may require another iteration with programming and budgeting if an appropriate fallback cannot be found with the given resources, and so on. Throughout this iterative process, MS&A provides the bond that enables the results of decisions made at each stage to be projected and used throughout. The role of the "consumer," as defined in Chapter 1 in the section on the role of MS&A, is key throughout this iteration.

In practice, the four processes are not well integrated in DoD. To address that situation, the 2006 *Quadrennial Defense Review* (Rumsfeld, 2006, p. 66) indicates that DoD is launching several initiatives to "integrate the processes that define needed capabilities, identify solutions and allocate resources to acquire them" and will "reach investment decisions through collaboration among the joint warfighter, acquisition and resource communities." To some degree, MS&A applications do currently support this integration. For example, mission analyses examine the levels of mission capability that are achievable for given resource expenditure, and campaign analyses examine trade-offs among overall components of force capability at given resource levels. However, the MS&A community can be a key partner in helping DoD achieve better and broader integration of its processes. To do so, MS&A must address a number of new challenges for resource allocation, planning, training, and operations.

Capabilities-Based Planning

A major challenge identified in the *Transformation Planning Guidance* (DoD, 2003) is the ability to move to capabilities-based planning (CBP). CBP stresses the development and honing of capabilities that can be applied to a wide range of tasks and circumstances. DoD is also moving toward an analogous capability for operations called adaptive planning, in which war plans are developed so as to anticipate and then support significant changes in those plans as circumstances change. The MS&A enterprise has an important role to play in CBP, but it must develop new capabilities in order to do that.

Network-Centric Warfare

Another major consideration for planning, training, and operations is network-centric operations. Networking is ubiquitous in all DoD activities, whether for peacetime planning or war itself. It is often referred to as network-centric thinking or network-centric operations—see, for example, Alberts and Hayes (2003), Alberts et al. (1999), and NRC (2000). By network-centric operations is meant military operations that are enabled by the networking of the force. Network-centric warfare (NCW) represents a powerful set of warfighting concepts and associated military capabilities that allow warfighters to take full advantage of all available information and bring all available assets to bear in a rapid and flexible manner.

The tenets of NCW are these:

- A robustly networked force improves information sharing.
- Information sharing enhances the quality of information and situational awareness.
- Shared situational awareness enables collaboration and self-synchronization and enhances sustainability and speed of command.
- These, in turn, dramatically increase mission effectiveness.

While there is considerable enthusiasm for the potential benefits of network-centric operations, practical experience is limited and the fundamental scientific base has yet to be built. As discussed in Chapter 3, there is an urgent need for research on how best to represent network-centric operations in models and simulations and how best to use MS&A in support of network-centric operations.

Reconstruction and Stabilization

Reconstruction and stabilization (R&S) is an important and often neglected phase of conflict and poses a new challenge for DoD. This period formally begins when (in the U.S. case) allied forces have satisfied military objectives to the point that sovereignty can be legitimately claimed. The R&S phase does not actually begin here, however, since planning and preparation for this phase must begin far in advance of execution. Successful reconstruction and stabilization depend significantly on successful control of many nonmilitary factors, from the provision of civil infrastructure to the management of political insurgency. Control implies more than the military enforcement of martial law; it requires integrated cooperation from foreign national agencies and from multiple agencies within the U.S. government.

The time frame for accomplishing R&S has fundamentally changed. Today, U.S. forces are so proficient at conducting the military phases of conflict that precious little time is afforded them along the way to learn about the civilian fabric. In Operation Iraqi Freedom, for example, allied forces were sovereign barely a week after military operations were initiated. The R&S efforts of the military were taken by surprise.

Two observations are key to the future of R&S. First, the roles and responsibilities of all the components must be defined and refined in advance, despite limited time available for preparing and conducting what-if exercises. Second, training and rehearsal capabilities are needed to support the activities required for R&S. The most promising technology-based approach to carrying out this task involves MS&A. Such capability would allow players from the various agencies to interact with one another, and it would provide a means by which planning, training, and evaluation could be standardized across the services in an era of evolving coordination.

There are several modeling issues for R&S. What data do we need to assess regional stability? How do we use these data to distinguish between stable and unstable regions? How do we merge them with existing data and share everything with all of the interested, possibly nonmilitary parties? Can we quantify and measure changes in stability as data or parameters change? Finally, do we have all of the MS&A tools that we need?

The problem here is that MS&A capabilities to aid planners in reconstruction and stabilization are sparse. There certainly are models that are useful for aspects of R&S. For instance, engineering models used to determine quantities of steel or concrete that will be required for the civil engineering components of R&S are evidently useful, but they do not help in representing cultural, political (perhaps tribal), or other human considerations that are organic to R&S. There are attempts by the services (e.g., the Army's Training, Doctrine and Analysis Command's analysis center) and industry (e.g., Alion's SEAS simulation product) to enhance current capabilities or to broaden them to satisfy some of the objectives that are assumed to be important for R&S MS&A. But there appears to be no product or near-term MS&A capability for planning or evaluating alternative courses for the vitally important R&S mission. On the other hand, the technological enablers in Chapter 3 (interoperability, composability, etc.) and the anticipated progress they are making suggest that MS&A for R&S is a distinct possibility. Within the government, DoD is arguably more technologically advanced in the relevant MS&A than other agencies involved in R&S and would be relied on to lead development of the technology that could be used by those agencies.

Diplomatic, Intelligence, Military, and Economic (DIME) Options

Because DoD is increasingly called on, as part of its evolving mission, to model the social and cultural aspects of conflict, it has an associated interest in developing models that go beyond the capabilities needed for R&S and take into account the entire range of political, military, economic, social, information, and infrastructure (PMESII) factors that might stem from, or guide, DoD decisions. This broadened range of DoD interests is consistent with the emerging viewpoint that DoD should consider an entire range of diplomatic, intelligence, military, and economic options when planning how to meet national goals. DIME is often used to designate a space of actions or options, whereas PMESII is used to describe a range of effects to be considered. Both stress the factors that contribute to and define the entire sociocultural environment surrounding the conflict. This vastly broader decision space puts heavy demands on M&S and the analysis that connects MS&A to decision makers. Ideally, DoD would like to develop the capability for modeling the entire range of factors and consequences that affect decisions, and in a way that enables easy evaluation of what-if scenarios and makes the assumptions and sensitivities readily visible. MS&A provides the only means for decision makers to gain experience and intuition about situations that have not occurred but that must be considered.

A general-purpose MS&A tool kit, one that would allow examination of even a small portion of the entire DIME space in any sort of integrated way, does not exist. For instance,

- There is no evidence that the PMESII factors completely define what is meant by the sociocultural environment.
- There are no common, agreed-on ways of defining, representing, modeling, or measuring these factors.
- There is no single unified social science theory that covers all (or even a major subset) of these factors.
- The state of an actor or country in terms of these factors changes over time.
- These factors interact in complex ways to impact the behavior of U.S. and adversarial forces at all levels, from the individual to the nation state.

From an MS&A perspective this means that, with respect to these factors, models at the current state of the art are incomplete. With rare exceptions, they are neither multitheoretical, multilevel, nor high-dimensional, nor do they deal with complex change dynamics. All of these attributes do not need to be present simultaneously for a model to be useful in MS&A, but more complexity than exists is needed for the problems cited above. Because there are no agreed-on definitions and representations, standard procedures for data collection do not yet exist. This lack of standard procedures, coupled with the inherent complexity and dynamics of the sociocultural domain, has two implications:

- The data needed to instantiate, tune, and validate these models are often nonexistent or at least woefully incomplete.
- The notion of validating models or at least establishing the conditions for their credible and responsible use must be completely rethought.

The committee addresses both issues in Chapter 3, in the subsection "Expanded Concepts of Validation."

Effects-Based Operations and Effects-Based Planning

Effects-based operations (EBO) is the name currently given to operations designed to achieve desired outcomes or "effects" through the synergistic, multiplicative, and cumulative application of the full range of military and nonmilitary capabilities at the tactical, operation, and strategic levels. The public literature on EBO includes Deptula (2001), Davis (2001), Smith (2006), and McCrabb (2005).[1]

Effects-based planning (EBP) is the staff process to work out the causal relationships and rationale for attacking various targets to support EBO. EBP balances that understanding with the known capabilities and risks. It is results-based as opposed to attrition-based, and it is much more specific than many classical expressions of a commander's intent in describing the assumed linkages of actions to objects, often

[1]Other information in this section comes from Maj Gen Robert Elder, Air War College, and Maj Gen Bentley Rayburn, U.S. Air Force, Briefings to the Conference on Effects Based Operations on January 31 and February 1, 2006.

through a sequence of intermediate effects. EBP closely mirrors the current joint planning process, yet focuses on the linkage of actions to effects to objectives. EBP changes the way we view the enemy, ourselves, and what is included and emphasized in the planning process.

EBO poses substantial challenges for MS&A because of (1) the uncertainties about the actual effects of particular actions; (2) the inevitable learning and adaptation resorted to by adversaries, third countries, and other groups; and (3) the many options, influences, and uncertainties that must be modeled and tracked. As the committee emphasizes in this report, EBP cannot depend on accurate prediction and must rely on a combination of reasoned initial actions, observation, and rapid adaptation. To do otherwise, to depend upon predictability, is a recipe for failure. In planning, a premium is therefore placed on finding flexible, adaptive, and robust (FAR) strategies that anticipate and facilitate adaptation, and MS&A methods must support those strategies.

Developing and analyzing alternative courses of action requires models that can represent complexity, adapt to situations as they arise, and explicitly account for the systems nature of most operations. As our MS&A capabilities improve, we are more able to represent some of the complexity of those defense systems that reflect these properties. These are described as complex adaptive systems (CAS).[2] Models that incorporate this view also should try to represent potential adaptation of all parties, random developments that always afflict real warfare, and other factors and processes that relate to political, military, and economic actions.

The emphasis on complexity must be reflected in adaptive models rather than the mostly scripted models that have been the mainstay of MS&A for many years. One promising approach for achieving adaptive models is agent-based modeling, although in forms beyond those taken by the most current work. Other modeling tools will be needed to incorporate human gaming and man-machine interaction, since it is commonly the case that expert human teams are better than models for suggesting innovative tactics and taking an integrative view. Over time, sophisticated models/agents might be developed that can substitute, to some degree, for such human teams, perhaps after such teams have first explored the concepts via games. The R&D needed for this is discussed in Chapter 3.

Complexity also requires a more thorough understanding of, and procedures for, assessing and managing the set of relations that connect various PMESII entities. Tools that employ link analysis, social network analysis, and data mining can be a move in this direction. However, they are currently limited by issues of scalability, difficulties in dealing with missing data, massive data entry requirements, and attention to only one or two relations at a time. They are also hampered by a lack of shareable ontologies, an unwillingness to use text-mining techniques, and legal, policy, and control issues that arise in sharing information.

Another dramatic feature of the future landscape for MS&A will be a substantial merging of MS&A into command and control systems. This is already happening as models used to monitor forces during execution and to conduct high-fidelity mission rehearsal become embedded in operational command-and-control systems. This trend will continue, bringing with it demands for MS&A that can reflect and routinely react well to real-time information that comes into the system during campaign execution. Continuous adaptation of plans will become more nearly routine. One simple current example is the launching of aircraft on missions before their targets are known. During the course of their mission, personnel may be directed to support a Special Forces team in trouble or the maneuver of an Army or Marine unit, to destroy time-sensitive targets that pop up from hiding, or to reattack fixed targets that were not adequately suppressed or destroyed by earlier attacks. Such on-the-spot adaptations may occur on a time scale of minutes. Representing such capabilities well requires high-fidelity models of a sort that were once associated more with training and exercises than with analysis and execution.

NEW TECHNOLOGICAL LANDSCAPE

Although a detailed examination of the new technologies that confront the military is beyond the committee's scope, future MS&A must be prepared to deal with them. Here the committee briefly surveys the most important aspects of the changing technological landscape.

Large Integrated, Interdependent Systems

Historically, the services and, often, elements within them have operated with a large amount of autonomy. Recently there has been a shift to interoperable elements, with the goal of migrating to a more efficient set of mutually supportive capabilities without inappropriate redundancies. The legacy of rigidly defined systems ("stovepipes") and their models is being rendered obsolete by emerging component-based systems. At the most abstract level, an example of this is the Joint Task Force (JTF) concept of a basic computer architecture that pulls together various components to provide the capabilities required for a particular mission.

Some of these JTFs are larger than any single service could provide, encompassing capabilities beyond those organic to a single service, with information as the interchange mechanism. This is the basic, overly simplified concept of NCW. In essence, the information interchange mechanism is the middleware layer and the interacting systems are the components. In some instances, overall system capabilities

[2]Some modelers use the term CAS to refer to the class of models, but in this report the committee uses the descriptor to refer to the real systems that those models seek to represent.

will be much more than the sum of the parts. This is familiar in the civilian world, where networking and distributed and collaborative activities are so ubiquitous.

The deployment of these integrated and interdependent systems has rendered many of the existing stovepiped models obsolete for representing emerging capabilities. To reflect the new military capabilities, the models need to implement a component-based architecture with a defined set of middleware similar to these information systems. Several years ago, the Defense Modeling and Simulation Office tried to advance this concept for DoD as the High-Level Architecture (HLA) was being implemented. Although the need was clear, the concept was valid, and implementation was effective in many respects, some technical and managerial missteps occurred. For example, the original implementation of HLA was not robust enough to permit dynamic behavior that had not been preplanned in a simulation, thereby limiting applicability. Also, overzealous attempts to enforce the use of HLA significantly impaired some experimentation, such as that in preparation for Millennium Challenge 2002. Those promoting the newer Test and Training Enabling Architecture have taken a more focused and voluntary approach to implementation of a standard middleware layer, and the large community engaged in distributed simulation currently uses an assortment of protocols, including HLA.

System of Systems

System of systems (SoS)[3] engineering deals with planning, analyzing, organizing, and integrating the capabilities of a mix of new and legacy systems into a new system whose capability is greater than the individual sum of its parts. SoS engineering is supposed to provide a comprehensive, collaborative, multidisciplinary, iterative, and concurrent technical management process encompassing the entire system life cycle, from the identification of systems capabilities through coordination of the development and integration of the parts, sustainment of the system, and system disposal.

What makes SoSs novel is that rather than being single monolithic structures, they are composed of multiple, autonomous, interacting, and interoperable systems. Thus, SoSs tend to be larger and more complex than the legacy systems they replace.

The movement to these large SoSs presents challenges for MS&A centered on scope and flexibility. Whereas previously system components could be narrowly defined and easily modeled, they now function more broadly with more complex models. SoSs are, by their nature, dynamic collections providing different capabilities at different times. Indeed, SoSs are expected to prove capable in circumstances that were not anticipated in any detail and were therefore not expressible in clear requirements of a classic sort. The models (and analyses) of these systems need to be able to adapt to requirements-driven changes and to circumstance-dependent demands in a timely and accurate manner.

As can happen in any large complex system, the interaction of the parts of an SoS may cause the system to behave in a manner that was not planned. This emergent behavior is something that must be captured in the model so that it is not first observed during actual combat. The issue of bandwidth in a service-oriented architecture is a classic example—the performance measures of traffic are not simply a linear function of the individual performances. Relations between network loading, capacity, and performance must be modeled and analyzed before being tested under fire.

Embedded Systems

A number of MS&A issues deal with the difficulties of designing and modeling embedded systems. Many of these issues concern real-time simulations coupled to embedded devices (hardware-in-the-loop) and modeling dynamical systems. A further challenge is hybrid modeling, a combination of modeling continuous physical devices and discrete computational devices (most frequently discrete controllers). Progress has been made in this area of research over the last two decades. Another challenge is network modeling. As will be discussed in Chapter 3, many serious analytic challenges remain in order to address the size and complexity of modern networks.

Unmanned Systems

According to current plans, within the next 20 years almost one quarter of the entities within the battlespace will be unmanned. Given the state of the art, this in an extremely aggressive goal. What makes the goal even more challenging is the almost complete lack of cognitive infrastructure for understanding, much less modeling, command-and-control aspects of the training, mission rehearsal, and conduct of operations of such devices. Unmanned systems have a greater reliance on models than manned systems because the actions of the former are not as amenable to human intervention. Moreover, robotic entities will likely be deployed alongside soldiers as part of mixed-initiative teams. There is a clear need to understand how such teams could operate and how command and control can be exercised. As the technology for unmanned vehicles (UxVs) advances, these systems will become more prevalent in the battlespace. Unmanned aerial vehicles (UAVs) are currently in use for both reconnaissance and attack teleoperation. Unmanned ground vehicles (UGVs) are being used for explosive ordnance disposal. In these cases, the UxVs are controlled via a communications link to an operator (sometimes more than one) who controls the vehicle. While this takes the human

[3]DoD guidance on SoS engineering can be found at http://www.deskbook.osd.mil/dag/Guidebook/IG_c4.2.6.asp.

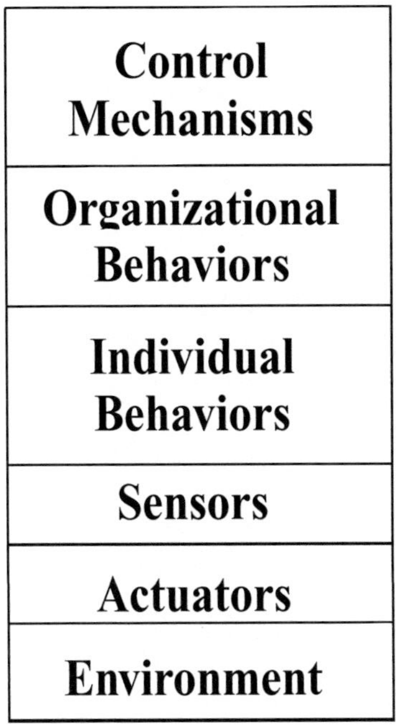

FIGURE 2.2 Controlling a system requires knowledge of the environment and control of all of the layers shown here.

out of harm's way, it does nothing to reduce manpower needs.

For that, a degree of autonomy will have to be embedded in the UxV. Assuming this can be done, questions arise that can only be answered by well-designed simulations: How will behaviors be developed? How is command and control going to be exercised? And most important, How is this new capability going to fit in with the organizational structure?

In the behavioral stack shown in Figure 2.2, the bottom three levels depend on the specific implementation, but the top three can be implemented on an abstract model of the platform. With this in mind, the behaviors used in the simulation might well be transferred to the actual UxV. This would allow the tools that are used for planning and mission rehearsal to be used for execution as well.

DoD is facing a new environment and new adversarial challenges. It will employ new technology and new management structures to deal with these challenges, with further advances already being planned. The current legacy systems of MS&A are insufficient to deal with these rapid changes. They deal with scripted, nonadaptive scenarios, take no notice of cultural factors in reconstruction, and do not recognize the network-centricity now prevalent. As part of this vision, agility in model development (probably through greater use of disposable and reusable models) is needed to reduce the cost of models. In the remainder of this report, the committee identifies steps DoD must take to realize MS&A that can meet these challenges.

ESTIMATING RETURN ON INVESTMENT

It is logical to ask why DoD should invest in MS&A. What is the benefit of that investment? A simple answer is that there is often no real alternative, because all military activity short of actual battlefield operations is, by necessity, a simulation, whether on an instrumented training range or in the CPU of a computer. But there are other reasons to invest in MS&A. Business models and military wisdom alike make it clear that MS&A provides methods and opportunities that realize a very positive return on investment (ROI). The ROI for any program can be evaluated in quantitative terms and, no less importantly, in qualitative terms, although such evaluation can be notoriously difficult. MS&A provides clearly exploitable ROIs from both perspectives.

MS&A is routinely used to explore factors such as required end-strengths, retention levels, the ability to prevail in multiple simultaneous regional conflicts, and other aspects of military operations, including combat. Evaluating the ROI of MS&A in such roles involves weighing the costs against the benefits. Costs include the construction, operation, and maintenance of simulation capabilities. Benefits include time savings, improved training, and safety. Various business-based models, including discount rate, return on investment, and net present value, are used to support the decisions that are made.

There are many concrete examples of ROI evaluations that have demonstrated cost savings or cost avoidances in the domains of training, acquisition, and force analysis. Examples come from all the services. Since the Navy has subsurface, surface, air, and land components, it is an appropriate service to use as an example. By one estimate, the Navy is expected to save approximately $130 million by using applications such as the following:

- *AIM-9X.* Flyout simulations (as opposed to live-fire exercises) developed for this air-to-air missile upgrade program will be used through FY09, resulting in more cost-effective test and evaluation.
- *Distributed mission trainers.* Joint, interoperable aviation simulations could provide savings of approximately $7 million per device over a 10-year period, according to Naval Aviation.
- *Joint semiautomated forces (JSAF).* Entity-level training simulation provides an alternative to the Navy's war gaming system as well as to related constructive simulation systems and provides joint interplay capability for sea-based and shore-based combatants through the deployment of standard simulation communication protocols. JSAF is the simulation model of choice for the Joint Forces Command, Joint Warfighting Center.

The quantitative ROIs provided by the above examples

are compelling. But there are other analyses of ROIs that are viewed as qualitative. For example,

- *Virtual at-sea training (VAST).* This M&S system provides the ability to conduct simulated multithreat scenarios at sea using a tactical scenario distributed from a shore-based combat simulation center. This simulation capability provides the Navy/Marine Corps with an alternative to the live-fire bombardment of training ranges, thereby eliminating a significant political obstacle. VAST transitioned to the fleet in FY04 and has since been used to qualify more than 25 ships before deployment. It had a triple ROI: (1) it solved a severe readiness problem caused by the loss of the Vieques training range in Puerto Rico, (2) it reduced by an estimated one order of magnitude the costs of naval surface fire support qualification by substantially reducing the costs of steaming to achieve qualification, and (3) it helped maintain readiness in ship-to-shore fires while in transit or on station. The effort led to the technological enablement of other significant capabilities to maximize fleet readiness while steaming to and in objective locations at sea using Battle Force Tactical Training technology modified for at-sea use.
- *USS Cowpens Combat Information Center Team Training (CICTT).* The quantitative ROI for one exercise alone was significant. CICTT saved 4,000 barrels of fuel, reduced required military travel costs, and reduced in-port costs. But the qualitative results were arguably more important. According to the Navy, training was improved significantly because the crew was better rested and able to focus on learning. Training objectives were completed in much less time than the previous work-up cycle, specifically in achieving Emergency Surge Proficiency Levels I and II.

Much can be and is said by the services about the qualitative value of MS&A. Prominent among the positive attributes of advanced simulation are flexibility and control. Globally networked simulators, some manned and others semiautomated, provide fidelity-managed scenarios for training and for actual combat rehearsal. Some of the advantages of this technology include the ability to stop a scenario, provide feedback, and reengage the mission. This capability can involve hundreds of manned players and thousands of behaviorally believable computer-generated actors. Scenarios can be varied on the fly and tactics can be trained at the individual, small group, or aggregate level. Tactics can also be changed in real time. Sentient red forces, played by tactically proficient coalition adversaries, provide instantaneous knowledge of results on the highly controllable, simulated battlefield. It is extremely difficult to quantify this advantage of MS&A. Both quantitative and qualitative ROI assessments clearly indicate the contribution of MS&A to the nation's military objectives.

Program managers and military commanders, all of whom must contend with budgets, are finding clear benefits in MS&A investment. The ability to quantify these benefits will be very important. By and large, military communities have traditionally resisted simulation in favor of live-fire training or evaluation, but this tendency will diminish as the technical quality of simulation improves. Many aspects of military operations—for example, the implications of ubiquitous networking, the implications of different types and degrees of information, and the potential political, social, and economic consequences of alternative courses of action—are not yet well understood, so M&S does not yet represent them well. Although much is known about counterinsurgency, and even about terrorism, techniques by which M&S can codify or apply that knowledge have not been developed. Although the past successes of M&S, partially enumerated above, support further development, quantitative justification would reinforce that support.

REFERENCES

Alberts, D.S., J.J. Gartska, and F.P. Stein. 1999. *Network Centric Warfare—Developing and Leveraging Information Superiority.* 2nd ed. Washington, D.C.: Department of Defense, Command and Control Research Program.

Alberts, David S., and Richard E. Hayes. 2003. *Power to the Edge: Command and Control in the Information Age.* Springfield, Va.: EBR Inc. Available at http://www.dodccrp.org/publications/pdf/Alberts_Power.pdf.

Davis, Paul K. 2001. *Effects-Based Operations: A Grand Challenge for the Analytical Community.* Santa Monica, Calif.: RAND.

Department of Defense (DoD). 2003. *Transformation Planning Guidance.*

Deptula, David. 2001. *Effects-Based Operations: Changes in the Nature of Warfare.* Arlington, Va.: Aerospace Education Foundation.

McCrabb, Maris. 2005. *Incorporating Effects into Military Operations*, Air Force Research Laboratory Report AFRL-IF-RS-TR-2005-237.

National Research Council (NRC). 2000. *Network-Centric Naval Forces: A Transition Strategy for Enhancing Operational Capabilities.* Washington, D.C.: National Academy Press.

Rumsfeld, Donald. 2006. *Quadrennial Defense Review.* Department of Defense.

Smith, Edward A. 2006. *Complexity, Networking, & Effects-Based Approaches to Operations.* Washington, D.C.: Command and Control Research Program, Office of the Assistant Secretary of Defense for Networks and Information Integration. Available at http://www.dodccrp.org/html2/pubs.html.

3

New Challenges and Directions for MS&A

CAPABILITIES NEEDED FOR DEFENSE MS&A

In Chapter 2, the committee identified three overarching themes that must be reflected in DoD's future development of MS&A: networking, adaptability, and embedded systems. These themes lead the committee to make three recommendations, which will be expanded on in this chapter:

Recommendation 1: DoD should give priority to developing flexible, adaptive, and robust MS&A methods for evaluating military strategies.

Recommendation 2: DoD should ensure that the basic architecture of MS&A systems reflects modern concepts of network-centric warfare.

Recommendation 3: DoD should give special emphasis to the development of MS&A capabilities that are needed within embedded systems.

These recommendations give rise to a set of functional needs. Because the new challenges in defense planning are dominated by uncertainty, solutions should emphasize strategies that are flexible, adaptive, and robust (FAR). That, in turn, requires an approach to MS&A that can help identify candidate FAR strategies and evaluate them. Adaptability is not a characteristic of most legacy systems, which include scripted (predetermined) data entities, strategies, tactics, and behaviors.

To elaborate, if DoD strategies and programs are conceived with branches and other features designed to cope with uncertainty, then the evaluation of options requires models that generate the realistic dynamic circumstances with which the strategic options will have to deal. Such models will need to reflect the learning, adaptive, and sometimes random behaviors of individual groups. They will also need to reflect the possibility of structural changes in the system as coalitions form or dissolve, key leaders emerge or disappear, and physical events change the realities of, say, geography or access. DoD can no longer evaluate its strategies with models conceived in a paradigm of well-defined closed systems. Adaptation can be achieved by drawing on methods derived from operations research, game theory, control theory, and agent-based modeling (see the subsection "Other Methods for Representing Adaptive Systems" in this chapter).

Another functional need is to ensure that people are employed effectively in the use of MS&A. One lesson learned over and over is that people are exceedingly capable when dealing with uncertainty or innovative concepts, or integrating across boundaries such as those associated with DIME and PMESII—indeed, often much more capable than traditional models and simulations. This superiority of human beings is so clear that, in some situations, gaming is a preferred method for operational planners and strategic planners. Gaming, however, has many limitations, including the potential for missing constraints imposed by the physics of the situation or by the real-world capabilities of systems. Further, humans have only limited capability to deal with complex nonlinear phenomena; they may be very creative and sometimes find novel solutions, but rigorous thinking amidst complexity is difficult. Furthermore, attempts to be rigorous often oversimplify the problem and obscure possibilities that are important, such as the diverse reactions of an opponent to one's own actions. It follows that MS&A should be reconceived broadly to include human gaming and interactive M&S for originality and insight. This applies whether MS&A is used for the support of strategic planning or for training, operations, or acquisition. Although neither gaming nor interactive M&S is rigorous, they have demonstrated the ability to reveal important factors and possibilities often missed by individual analysts or decision makers.

Representing Complex, Dynamic, and Adaptive Systems

Previous generations of MS&A were developed largely with perspectives that we now associate with idealized problems and mass-dominated forms of warfare emphasizing attrition and, sometimes, maneuver. This style reflected the educational background of the model builders and the experience of the United States in two world wars. These were "industrial wars," and the U.S. style in war was reasonably described as winning through sheer overwhelming force with large military forces and prodigious quantities of aircraft, ships, and tanks (Weigley, 1973). In addition, the individual services fought separately and had clear sectors of responsibility, which simplified deconfliction. Finally, the U.S. military thought about war as combat itself, with relatively little discussion of gray area activities before, during, and after combat.[1]

Over the last 15-20 years, the shortcomings of that approach to MS&A have increasingly been recognized, and the skyrocketing power of computing has made richer approaches possible. As far back as 1980, a new approach that would be, in today's terminology, more joint and more integrated with both political and military considerations was being considered by DoD. The Office of Net Assessment sponsored an ambitious undertaking along those lines at RAND. With the end of the cold war, the disappearance of the Soviet threat, and the temporary loss of interest in big models and games, the effort dissipated despite its successes, although leaving behind the improved Joint Integrated Contingency Model (JICM), which is now an important element of analysis in parts of DoD. The effort also stimulated a great deal of research into planning under uncertainty, which has contributed significantly to today's concepts of capabilities-based planning (CBP) and adaptive planning (the phrase used by DoD when referring to operations planning under uncertainty).

During those same 15-20 years, researchers in a number of scientific disciplines were making progress in modeling complex adaptive systems (CAS). Chemists, physicists, biologists, economists, engineers, and others nurtured and shaped CAS theory as a powerful way to look at much of what goes on in the real world. The approach to modeling CAS includes systems of interacting actors at different levels of organization, actors that have goal-seeking behavior and that can learn, adapt, and interact in ways that sometimes lead to higher-level phenomena (emergent behavior) whose character is not predictable by viewing the individual actors.[2] This extended earlier thinking about systems, such as the system dynamics methodology pioneered at MIT by Jay Forrester,[3] and added an emphasis on nonlinearities and an increased ability to predict events that were sensitive to initial and subsequent conditions.[4]

By recognizing this fundamental fact—that is, that the real world is a complex adaptive system—one's approach to MS&A changes substantially. Table 3.1 displays some of the changes. The first column applies to relatively simplistic views and models; the middle column applies to a few defense models of the 1980s, such as the RAND Strategy Assessment System (RSAS), Eagle, and the Navy Simulation System (NSS); and the last column arguably applies to the future. If the table is roughly correct, then the conclusion is inescapable that representing jointness, DIME/PMESII aspects, and asymmetric strategies requires modeling that is mindful of the paradigm of complex, adaptive, and dynamic systems. Moreover, that conclusion does not depend on whether particular methods of current mainstream CAS research prove enduring.

Table 3.1 might suggest an inexorable movement toward complexity—that is, a move from simple models to something inherently deep and detailed. There may, in some respects, be such a movement, but movement toward complexity is not what this report recommends. Instead, the committee sees the need for *families* of models and games varying greatly in level of detail and perspective. Perhaps most models should be relatively small, specialized, and readily understandable, with only a few models and simulations used for integrating concepts, as described in the last column of the table. The model-family concept is discussed further in the section titled "Promising Technical Approaches for Attaining the Needed MS&A Capabilities." The importance of relatively simple models is also discussed there and elsewhere in this chapter.

Features mentioned in the intermediate column were achieved to a significant degree in some 1980s-era systems, such as the RSAS, the Eagle, and the NSS and to some extent in the Joint Warfare System (JWARS) system currently being tested.

Representing Embedded Systems

A special challenge in representing complex systems arises from the need to develop MS&A capabilities that are part of embedded systems. In such cases, there might be feedback loops, dynamic incorporation of new data, and complex interactions between real (sensed) and simulated data. The state space becomes enormous, yet quality assurance must be high. There are many unresolved challenges as a result.

Although embedded MS&A capabilities have existed as an integral part of many systems, especially DoD systems,

[1]See Hughes (1989) for a good discussion of U.S. military modeling into the 1980s.

[2]See Waldrop (1992) for an excellent popular-level account of the Sante Fe Institute's early work. See also Holland (1995) for an excellent scientific discussion that can be understood without mathematics.

[3]For a more recent account of system dynamics, see the text by Sterman (2000).

[4]For a theoretical treatment of the connection of dynamical systems and simulation, see Nagel et al. (1997, 1999).

TABLE 3.1 Levels of Model Sophistication

Aspect	Simplistic	Intermediate	Advanced
View of war	Continuous "piston-driven" battles. Simple system depictions with with air, maritime, and ground components.	Maneuver leading to discrete battles with attrition and movement affected by material and qualitative considerations. Richer logistics and depiction of political-military systems, with some models representing decision makers' war plans.	Preparation, combat, and stabilization and reconstruction, substantial political and economic aspects. Allow big shifts of war trajectory as result of changes in leaders, coalitions, forces, or events. Logistics with just-in-time and responsive aspects.
Number of parties	Two sides, with allies folded into the appropriate side.	Plus some explicit modeling of third countries.	Plus nongovernment organizations and threats.
Decision making and strategies	Implicit in data or behavioral algorithms in specialized models.	Top-down decision models, political and military branches and adaptations.	Top-down, bottom-up, and distributed decision making and behavior at all levels, sometimes emergent.
Instruments	Physical attrition and targeting.	Plus some mechanisms for escalation, de-escalation, or or termination.	Plus nonkinetic attacks and mechanisms of coercion and dissuasion.
Attrition and targeting mechanisms	Difference equations with situational coefficients; direct physical destruction.	Plus per-sortie or or per-shot kills, breakthrough effects, and other embellishments.	Plus nonkinetic kill mechanisms and effects-based operations.
Nature of variables	Only objective variables, such as a side's firepower.	Plus soft variables such as a side's fighting effectiveness, affected by morale, leadership and other factors.	Plus soft variables such as nationalism, ethnic group association, and propensity for brutality and terrorism.
Command and control (including information assurance and intelligence)	Assumed perfect.	Plus specialized technical models of communications. Plans predetermine who does what to whom.	Plus network-centric concepts with, e.g., publish-subscribe architectures and capacity for self-synchronization.
Intended purpose of model runs	Emphasis on predictive modeling. Some sensitivity.	Multiscenario analysis with recognition of great uncertainties.	Exploratory analysis in search of strategies that are flexible, adaptive, and robust.

their number and diversity have greatly increased, from handheld devices to spaceships, and hence vary greatly in the constraints on computational power, memory management, and timing requirements. The projected use of unpiloted ground and air vehicles or swarms of vehicles that can respond autonomously to environmental changes or changes in enemy actions observed by onboard detectors is a prime example of the importance of embedded systems to future military capabilities. In order to effectively respond to changes, these vehicles need to autonomously project the effect of their courses of action. Despite the widely varying scales and capabilities of embedded systems, most embedded systems have the same impacts on MS&A.

Most prominently, in an embedded system, the MS&A has gone from being an offline computational capability for reasoning about the system to being an integral part of a performing system in an online, often real-time capability. This mode of operation for MS&A requires new types of traceability, new ways to provide checkpoints for the system so that it can back up to previous decisions and states within a rapidly changing circumstance, and new ways to evaluate and report partial results amid ongoing computation. It also requires new self-monitoring and self-analysis capabilities that provide, for example, a computationally reflective MS&A system that monitors its own state and acts on and modifies itself.

Because embedded systems must use the currently available data within a fixed amount of time and must therefore sometimes yield intermediate results, we will need advances in methods for evaluating the goodness of the current solution or the computational progress so far—that is, how far the current solution is from the optimal and how much additional value can be obtained by continuing computation.

M&S embedded within a command and control system exemplifies what is generally known as a dynamic, data-driven application system (DDDAS). Such systems are the target of a major program at the National Science Foundation.[5] In such a system, the embedded M&S is expected to control and guide a measurement process, determining when, where, and how to gather additional data. The embedded M&S must operate at both a global level—determining which systems to use to collect more data—and a local level—guiding particular systems as they gather measurements.

The vision of a DDDAS-style system also includes a second major goal, the incorporation of dynamic data inputs into an executing application in order to have the currently most accurate data available for models and other computational processes. This vision includes the ability of the system to accept and respond to dynamic inputs from live data sources that might include users, computational processes, archival data, or sensors.

[5]Information available online at http://www.dddas.org.

One of the hardest challenges in such systems is the reconciliation of the modeled world with a continuous stream of newly measured data. To the extent that this challenge can be met, the advantages are clear for a large number of current MS&A applications, especially in time-critical applications such as route planning. Embedded systems that incorporate simple forms of MS&A are already tackling this hard issue of reconciling and updating their models with newly sensed information. However, the state of the art requires a lot of hand-tuning, and there has been little analysis or validation of these early systems.

Representing Networking

Through most of the 1990s, DoD's models and simulations were largely developed with the technical aspects of command and control taken for granted or, at best, treated by separate communities, such as those dealing with communication issues or space surveillance. Within most combat models, command and control was largely assumed to work, except perhaps for parameters representing delays and hard-wired relationships allowing only some organizations to communicate with others. The mental picture of command and control was often point to point, and if one had a specific problem in mind, such as requiring the presence of a particular surveillance platform, that platform could be added with specific point-to-point links.

The modern concept of networking is quite different from this point-to-point perspective. Some of the capabilities enabled by the increasingly ubiquitous presence of networks and associated services are the following: (1) planners can draw information worldwide without knowing the specific locations where the information resides, (2) operating forces can obtain situational-awareness (sensor and intelligence) information without hard wiring sensor-to-weapon-platform links, and (3) flexible command and control and organizational relationships can be established to meet the needs of immediate contingencies and then to adapt as the situation evolves.

This type of networked world and the concomitant network-centric operation suggests the need for a new generation of MS&A having basic architectures in tune with modern-day concepts. Such architectures would probably be quite different from the prenetworking architecture, which had a simple overlay of particular platforms for sensing and communication. Some strides are being made, but the committee believes that no clear consensus yet exists for how network-centric operation should be represented in DoD's MS&A, either as retrofits to legacy models or in future models.

The need for this new MS&A capability is critical. It is required to help develop operational concepts and force structures for U.S. and coalition military forces, allowing them to meet and adapt to the new threats facing them, particularly in light of uncertainty about where and in what manner threats will arise. Furthermore, the DoD investment in developing network-centric capabilities is large—the expenditure estimated for the Global Information Grid (the emerging networking and information infrastructure) through 2011 is $34 billion (GAO, 2006), and analytical means are necessary to optimize this and later investments.

The following examples are representative of the current state of network-centric M&S for the major DoD process areas shown in Figure 2.1.[6]

- *Concept development and capabilities definition (example 1).* Agent-based models with simple agents (i.e., few rules governing their behavior) have demonstrated qualitative behaviors of network-centric operation (e.g., self-synchronization of force elements). As such, these models can serve as exploratory tools, but they do not provide the quantitatively supportable results needed for more detailed analysis.
- *Concept development and capabilities definition (example 2).* Numerous experiments involving live and simulated forces have been conducted to explore network-centric operational concepts. While a variety of models and simulations support these experiments, they themselves do not typically embody network-centric concepts—rather, the network-centric behavior is achieved by connecting the individual models and simulations over physical networks.
- *Programming and budgeting.* Traditional models (CASTFOREM, VIC, NSS, and others)[7] have been used by the military services in making major resource allocation decisions affecting the development of network-centric capabilities (e.g., for the Army's Future Combat System). While the services indicated they had made significant progress in the course of these analyses, these activities were assessed as being only initial efforts in addressing network-centric operations (MORS, 2004).
- *Acquisition and engineering.* Network Warfare Simulation (NETWARS) is the Joint Staff's standard model for measuring and assessing information flow through existing and planned military communications networks. It is intended for analyzing performance resulting from behavior at the physical layer through the network layer, and it provides an extensive capability for doing so. However, its focus on the lower layers of the network precludes it from modeling important technical factors such as information assurance (beyond encryption), higher-layer services (e.g., discovery and collaboration), and ad hoc entry to and exit from networks that are critical to envisioned modes of network-centric operation.
- *Training and operations.* Agent-based models coupled with the techniques of dynamic network analysis have been applied to address operational problems confronting combatant commands. Examples include assessing and improving the organizational behavior of command and control staffs and characterizing the network structure of terrorist threats and their surrounding social environment. While these examples illustrate how the underlying technologies contribute to operational use, much basic research must be done and empirical data collected for a broad, robust application of the technologies.

In summary, there is activity in each of the four DoD process areas shown in Figure 2.1, but in none of them is there yet a broad conceptual basis for addressing the problems of the area.

Given the need for MS&A to represent network-centric operations and the current state of such representation, as indicated by the examples above, the committee recommends as follows:

Recommendation 4: DoD should establish a comprehensive and systematic approach for developing the MS&A capabilities to represent network-centric operations:

- **Enhance and sustain collaborations among the various parties developing network-centric MS&A capabilities.**

As it researched the examples given above, the committee found little evidence of significant interaction and cross-fertilization across the application communities associated with each of the examples. Such interaction is necessary to promote innovation and establish broadly based capabilities. The necessary collaboration might be facilitated by a DoD-sponsored series of workshops involving all the communities, leading to a substantive report synthesizing the views of the different communities and identifying opportunities for cross-fertilization.

- **Continue and extend the development of existing approaches to modeling network-centric operation.**

Since the basic architecture and functioning of traditional models reflect a prenetwork perspective on military operations, those models are not adequate for describing network-centric operation. Still, they cannot be abandoned at this time since no other means exist for supporting certain important quantitative analyses such as resource allocation. Agent-based models have shown some promise, and

[6]A major capabilities-based analysis of network and services infrastructure (the so-called Net-Centric Operational Environment) sponsored by the Joint Staff and being carried out by the RAND Corporation was planned for completion in May 2006. When completed, that analysis should provide a further example of the application of modeling and simulation to network-centric operation.

[7]CASTFOREM, Combined Arms Task Force Engagement Model; VIC, Vector in Commander; and NSS, Naval Simulation System.

TABLE 3.2 Important Directions Recommended in This Report for Advancing the Capabilities of Defense MS&A

Area	Recommended Topic	Page
Navigation of large state spaces	Exploratory analysis	23
	Multiresolution modeling	24
Representing complex adaptive systems	Optimization and agent-based models	25
	Multiagent systems	26
	Social behavioral networks	27
Fundamental scientific issues	Serious games	28
	Network science	29
	Embedded MS&A systems	29
	Expanded concepts of validation	30
Infrastructure needs	Composability	32
	Improved data collection for MS&A	32
	Visualization of high-dimensional data	35

further development is warranted. Attention should be given to the use of complex agents with sizable rule sets governing behavior to provide quantitative models and to the continued coupling of agent-based models with the techniques of dynamic network analysis (for a fuller discussion of dynamic network analysis, see the subsection "Social Behavioral Networks," later in this chapter, and see Appendix B). In contrast to bottom-up agent approaches, a top-down architectural approach to describing networked behavior is also desirable, but no good examples of such an approach exist yet.

- **Establish a new mathematical basis for models describing network-centric operation, drawing on an array of approaches, particularly complex, adaptive systems research.**

Just as new mathematics or the new application of existing mathematics has been necessary for advances in science, so too might a new network-based mathematical framework be necessary to realize appropriate models and simulations for network-centric operations. In general, research in the complex adaptive systems community could provide a basis for this framework. Some ideas along these lines have been put forward based on the mathematical structure of networks (Cares, 2006), and the methods underlying dynamic network analysis should also be applicable (Carley, 2003).

PROMISING TECHNICAL APPROACHES FOR ATTAINING THE NEEDED MS&A CAPABILITIES

To build the capabilities described in the preceding section, progress is needed in four areas of MS&A:

- Tools are needed for navigating intelligently through the very large state spaces characteristic of complex, dynamic networked systems. Some promising directions are described below, in the subsections devoted to exploratory analysis, multiresolution modeling, and families of models and games.
- Methods are needed for representing complex adaptive systems. There has been real progress in this direction through agent-based modeling, other means of representing adaptive systems, serious games, social behavior networks, and network science.
- Research must be stimulated to address fundamental questions that limit our ability to create scientifically sound and empirically grounded MS&A of the necessary complexity. There are many open questions about the analytical basis for such complex models, and validation is still more of an art than a science.
- New requirements for MS&A will need capabilities based on new infrastructure.

In the rest of this chapter, the committee recommends directions for advancing the capabilities of defense MS&A, as shown in Table 3.2. Many of these topics are already recognized within parts of DoD as logical steps for the evolution of MS&A, and relevant R&D exists or is beginning. Therefore, the table should not be taken to imply that these are new or unappreciated ideas; rather, it demonstrates an interwoven vision of the many important topics in the vanguard of defense MS&A and recommends that DoD and the MS&A community adopt this holistic view for advancing DoD's capabilities to address the needs identified in Chapter 2.

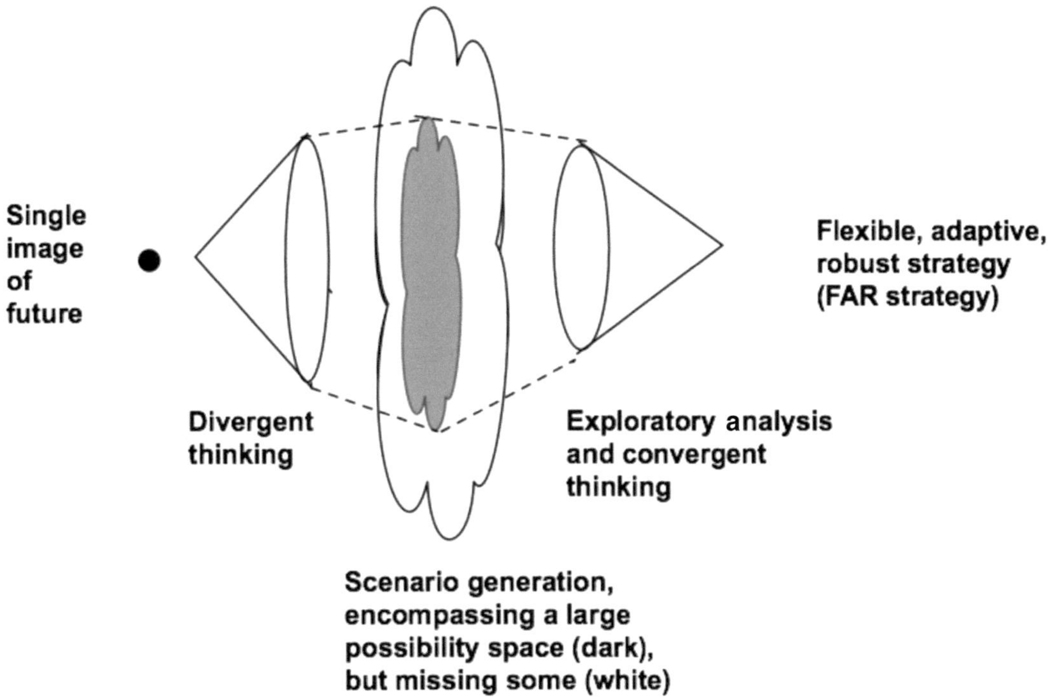

FIGURE 3.1 Divergent and convergent thinking in search of FAR strategies.

Exploratory Analysis

MS&A is needed to assess, among other things, whether an option under consideration is flexible, adaptive, and robust (FAR). The option must be evaluated across as broad a space of scenarios as can be conceived. In addition to having suitable models and games for such an evaluation, one must also have methods for generating cases throughout the space, be able to use M&S to characterize results from each case, and then be able to make sense of the results. Figure 3.1 provides a schematic depiction of the kinds of thinking involved in developing and evaluating FAR strategies

The sheer dimensionality of the space of possibilities can make FAR strategy development a daunting task. A rich collection of methods for exploratory analysis has developed over the last decade or so.[8] These include methods for structuring the initial possibility space, divergent thinking to expand notions about what is possible (to include, for example, the possibility that people and groups will adapt or that events that change the very structure of the system will occur), generating simulations, visualizing outcomes, and applying a variety of tools—some of them derived from data mining or cluster analysis, some from the artificial intelligence and statistics communities, and some from outside-the-box gaming or brainstorming—and then finally converging toward reasonable depictions of alternative strategies and assessment of their merits. Exploratory analysis is very different from traditional sensitivity analysis. Whereas sensitivity analysis typically examines how the outputs of a complex model or simulation change when parameters and inputs deviate slightly from nominal values, exploratory analysis typically attempts much broader coverage of the space of parameters and inputs from much simpler quick-and-dirty models. Different versions of exploratory analysis apply to planning and programming system development, and operations, and the difference between the versions is large.[9,10]

Some of the challenges associated with exploratory analysis are deeply technical while others have to do with how best to structure collaborative analyses involving both hu-

[8]Exploratory analysis is discussed in a larger context in Davis et al. (2005). An early reference drawing upon a decade of work on multiscenario analysis is Davis (2003). For discussion of the related method of exploratory modeling in the context of long-range planning, see Lambert et al. (2003).

[9]Some DoD applications are discussed in Johnson et al. (2003).

[10]Exploratory analysis is not like sensitivity analysis as it is ordinarily practiced, varying one or two parameters at a time around some base case. It involves exploring the entire space implied by the domains of the model parameters. Because of massive uncertainties, exploratory analysis often rejects even the concept of best-estimate base case. Modern statistical analysis sometimes examines the entire space of a small, well-defined model but seldom looks into uncertainties about the structure of the model itself.

mans and machines and how best to summarize results for decision makers. While some decision makers prefer firm predictions and dislike uncertainty, many have a great interest in understanding uncertainty and how their course of action can both allow for opportunities that may arise and hedge against downside events. The issue becomes how to convey that kind of information clearly and accurately, a topic discussed further in Chapter 4.

Multiresolution Modeling and Families of Models and Games

A multiresolution model (MRM) is a model that can accept inputs and/or perform analyses at varying levels of resolution. True multiresolution modeling is very different from modifying a bottom-up model that requires high-resolution inputs to enable it to display aggregated (low-resolution) outputs. One motivation for MRM is the recognition that people need to reason at different levels of detail. At any given level of decision making, people do most of their reasoning with the natural variables of that level. In addition, they need to be able to zoom to the next more detailed level, so as to understand factors and phenomena underlying these higher-resolution features. Furthermore, they typically need to be able to summarize their reasoning to their own superiors, abstracting it into a form suitable for a lower-resolution level. The need for models at different levels of resolution cannot be addressed by simply applying sufficient computing power to do all modeling at the highest resolution. There are fundamental issues associated with aggregation and drill-down that must be understood and incorporated into models if multiresolution models are to provide effective support to decision makers.

To some extent, a given model can be designed so that it can be used at different levels of detail. However, because MRM models can become quite complex, at some point it becomes easier and clearer to have an integrated family of models. Sometimes it is adequate to have a family of models that were not designed in an integrated way but that are sufficiently consistent and sufficiently well understood to be able to inform one another. For example, if one has a trusted high-resolution model, it can be used to develop values, value ranges, or even probability distributions for the inputs to a higher-level, more-aggregated model of the family. Further, if one has a trusted low-resolution model, perhaps informed by solid empirical experience (including history), it can be used to inform higher-resolution models. Sometimes simple models reflect considerations such as the morale and fighting effectiveness of a nation's army, which have usually been assumed away in higher-resolution models.[11]

[11]See Dupuy (1987). Such insights are reflected in some models, such as JCIM and, more recently, JWARS.

Exploratory analysis is arguably best accomplished with a good aggregate-level model that can cover the entire possibility space clearly, albeit at low resolution. Such a model might have 6 to 10 variables. Understanding the outputs of the model over the entire space the values of those variables is quite feasible with modern tools. Further, one can then reason at that level of detail. If one does such a synoptic exploration and finds that only two or three of the variables are particularly important, then with MRM or a suitable family of models, one can zoom to higher resolution on those variables. This provides a straightforward, cognitively natural way of conducting exploratory analysis. In contrast, if one starts with a complex model built bottom-up, the model may have thousands of inputs (especially if one realizes that the individual items in the model's complex databases are all uncertain). Making sense of that model's behavior and finding abstracted insights can be exceedingly difficult and treacherous.[12] Thus, while MRM is not necessarily essential for exploratory analysis, it is a strong enabler.

Another enabler for exploratory analysis (again useful for other purposes as well) is having families of models, human games, experiments, and other sources of information (Davis, 2006). Figure 3.2 illustrates this by suggesting the strengths and weaknesses of some of the various instruments that can be brought to bear. Although the cell-by-cell evaluations depend on various assumptions and are only approximate, the story conveyed is valid. For example, relatively simple analytical models and programs (top left) are excellent for agility and breadth of work and for high-level decision support but poor for revealing underlying phenomenology. In contrast, detailed models, bottom-up, agent-based modeling (discussed later), human games, field experiments, and history can be very good at representing and studying phenomenology. Human games and man-machine gaming can also be particularly good for coming up with innovative concepts of operations, clever tactics, and new uses of technology. Strategic-level simulations can be excellent for integration, especially if they have adaptive decision models.

Recommendation 5: DoD's analytical organizations should take a portfolio approach to designing their analyses and supporting research, investing in a range of methods including diverse models, games, field experiments, and other ways to obtain information.

[12]In some cases, the detailed model can be exercised with a statistically designed experimental plan, and its outputs can be analyzed statistically. It may be that despite the model's complexity, only a few variables matter, or that only a few composite variables matter. In other cases, however, there are subtle and complex interactions among the variables that make statistical analysis either difficult to construct or hard to interpret. See Davis and Bigelow (2003) for discussion.

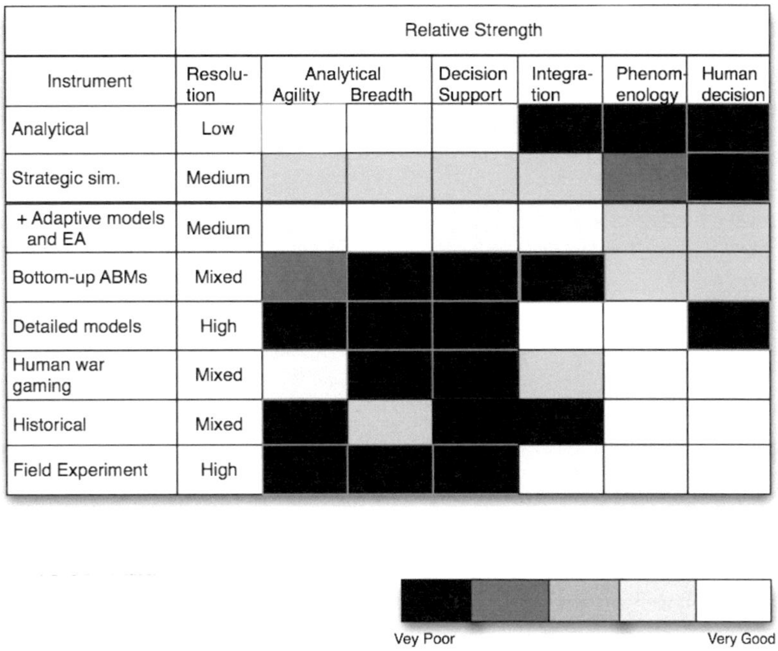

FIGURE 3.2 Relative strengths for MS&A models.

Such organizations should be cautious about (1) allowing the high-cost MS&A activities to use all of the investment resources, with no groups doing fast, simple, and nimble thinking; and (2) depending entirely on computer models, which may be unrealistic because they lack human involvement and often do not use real-world data, such as lessons-learned information from recent wars.

If exploratory analysis were used by analytical organizations in cooperation with other organizations, something else would probably happen: increased analytical structuring of human games and exercises. Games and exercises are rarely designed with the idea of building a consolidated knowledge base. But they could be, in which case human games would be tailored and analyzed accordingly, as would experiments, resulting in enhancements to a knowledge base and to models. For example, a theater-level model could have submodels to represent commanders' decision making. Those, in turn, could be made to address issues arising in human war games (as well as many other issues not arising in the games). In practical terms, this might involve building into the model aspects such as substantial decision delays except in circumstances of prior alert and prior authorization for rapid action. Decision delays might also be explicitly dependent upon the availability and quality of information that is not from satellites or aircraft but from U.S. national and regional intelligence, personal commander-to-commander conversations with allied officers, or special information from, say, a nongovernmental organization aware of circumstances on the ground. That is, human gaming could force MS&A to incorporate variables that are important in the real world of DIME/PMESII but not natural to those building traditional mathematics-based models.

Optimization and Agent-Based Modeling

It is often the case that mathematical optimization is needed as a component of a larger simulation or to establish performance bounds on the results of a simulation. While in some cases the simulations required for the challenges cited in Chapter 2 may be too large or complex to optimize in any formal sense and the uncertainties may diminish the utility of optimization, in other cases optimization techniques can be quite useful. It is important to realize that such techniques are available when needed. Many "purposive" military systems are currently modeled as a collection of smart, and arguably rational, decision-making agents that attempt to continuously improve some overall objective function. Although this paradigm has its defenders and critics in the modeling community when used to model some nonmilitary systems, it has been recently shown that the paradigm of self-optimizing via agents can be used to optimize a variety of general large-scale complex systems (Ghate et al., 2005). These need not be systems of antagonistic elements, as is often assumed for military analysis. For example, distributed and/or decentralized control architectures have been studied in artificial intelligence and robotics (Bui et al., 1998). The typical setting involves a group of agents that have (more or less) homogeneous capabilities, share a common objective, and, critically, have access to an ad hoc protocol set for a large number of contingencies and for coordination among themselves. Since these protocols assume assured bilateral agent communication, determining the best protocols to use becomes computationally intractable when the number of contingencies and/or the number of agents increases.

An alternative agent-optimizing approach, known as fictitious play (FP) (Brown, 1951; Robinson, 1951), addresses the challenge of optimizing black-box simulation models that cannot be expected to exhibit the kinds of regularity or convexity properties that conventional nonlinear optimization approaches demand (Bazaraa et al., 1993). The basic idea, inspired by game theory, is to animate the design components or controllable variables of a system by representing them as the decisions of intelligent, goal-seeking agents. These agents attempt to optimize their own selfish responses to an environment created by the behaviors of the other agents/components. This process can be viewed as an iterative game, with the components being players having identical interests—the overall performance of the system. Although in its infancy, this approach has been successfully applied to problems in the private sector (e.g., the joint optimization of plant-level production, capacity planning, and marketing decisions at one of the big three automakers) and the military (e.g., allocating resources and routing messages in mobile ad hoc networks and determining optimal ship-steering policies).

For example, in typical traffic routing in a dynamic network, an individual vehicle has origin and destination locations, an origin departure time, and a finite set of sequences (routes) of road segments joining the origin with the destination. Each vehicle's route traversal time is influenced by the traffic congestion on each link in the route during the time the vehicle is traveling along the link, so that the route traversal time depends on the choices of routes made by the other vehicles in the network. This system-optimal traffic assignment problem with flow-dependent costs has been studied extensively.[13] However, a crucial distinguishing characteristic of this problem—namely, dynamic, time-dependent congestion on the links in the network—has rarely been considered. In this version of the problem, numerical procedures for evaluating transit times are simulation based (down to the individual vehicle) and require significant computational effort. The task of finding the system-optimal routes is therefore inherently complex, and it is the subject of a great deal of research in the fields of intelligent transportation systems and sequential dynamic systems.[14] However, an FP algorithm has recently been proven successful in addressing this problem (Garcia et al., 2000; Lambert et al., 2005).

There are several general computational advantages to FP when trying to analyze models of military complexity:

- It can be applied to complex optimization problems with a black-box objective function lacking special structure.
- It updates all decision variables independently and in parallel, making the approach scalable in the number of variables (unlike other global optimization approaches, such as simulated annealing).
- Convergence to an optimal solution is ensured in the limit (Ghate et al., 2005).
- In many applications, only one evaluation of the objective function is needed per iteration.
- In practice, very fast convergence to high-quality solutions is observed.

Because of its general nature and its ability to optimize simulation-based models, FP warrants a thorough investigation. The committee has chosen to focus on FP as a particularly promising technique for optimization of highly complex, nonlinear systems. There are, of course, other promising approaches that merit investigation. The broader point is that serious investigation is needed into theories and methods for optimizing the performance of complex, adaptive networked systems.

Other Methods for Representing Adaptive Systems

The use of multiagent systems (MASs) allows developers an often appealingly intuitive and straightforward way of incrementally developing complex systems in a distributed and locally adaptive fashion. These are explored in the subsection after next. However, MASs are not the only way to model adaptive systems, and it is important that DoD continue to actively use, research, and develop other methods for modeling adaptive systems, as well as ways to compare the benefits and limitations of different modeling methods across different classes of problems. Different modeling methods are called for because of variations in the depth and fidelity required for a given application area and because of implementation issues, including efficiency, development time, and the expected operation of different asynchronous and autonomous segments of the system.

Modeling methods that can represent adaptive behavior are needed because in many systems one cannot specify in advance all the conditions that could prevail and all the data that might be obtained. Furthermore, in many large, distributed real-time systems, no central decision-making element is fast enough to respond as needed to locally changing conditions.

The word "agent" has been used for models, however implemented, that can generate solutions in an adaptive manner. Examples include genetic or evolutionary programming methods for solving optimization problems, and game-theoretic, control-theoretic, or rule-based methods for solving decision problems. However, in order to gain the advantages of different MAS methods and other methods for representing

[13]See, for example, Potts and Oliver (1972).

[14]See, for example, Patek et al. (2001).

adaptive behaviors, it is important to distinguish among the different desirable characteristics of adaptability exhibited by different implementation strategies and to identify the applications for which each is most desirable. It should be noted that these heuristic methods may be superseded in the future by new theoretical developments in the mathematics of optimization or by increased computer power. This report has not focused on techniques of deterministic optimization.

When we talk about modeling methods that allow both local decision making and local individual history in a heterogeneous distributed environment, it is very natural to think in terms of agency. Agency can be implemented with any distributed, object-oriented modeling technique not just with those explicitly labeled "software agents." The ability to adapt to specific, time-dependent inputs can also be implemented with decision rules, knowledge bases, and logic-based programming methods.

In a well-characterized solution space, there are mathematical methods to combine or integrate local results, often described using sets of equations. Partial differential equation solvers and other synchronous computational methods are examples of important methods that may be usefully incorporated into an agent-based framework. That is, although equation solvers are sometimes viewed as antithetical to an agent-based solution, it is quite feasible to design an agent-based system in which some of the agents use inputs from their environment to construct equations that they then employ an equation-solver to solve and provide the results to other agents in the system. Similar remarks obviously apply to optimization, regression, and other methods often placed in opposition to agent-based systems.

Currently deployed adaptive systems use a wide range of strategies for adaptation. Some—for example, control systems for complex electromechanical devices and systems such as UAVs—choose among existing models of the environment in response to results of measurements. The best of these allow for some online model building to help react appropriately in the short term to unexpected behaviors in the environment (usually combined with a call for human intervention). In addition, any amount of self-modeling that can be usefully interpreted allows the system to examine its own capabilities and plan its activities much more effectively, including identifying and reacting more quickly when problems occur and even determining in advance when problems might occur.

Social Behavioral Networks

A behavioral model is a model of human activity in which individual or group behaviors are derived from the psychological or social aspects of individuals. Much progress has been made in recent years in this area, and it is of central importance to many questions addressed by MS&A across the DIME space. There are a number of approaches to behavioral modeling. Among these, the key computational approaches that are important from a DoD perspective are social network models and multiagent systems. In this section, the committee briefly describes both approaches and their variants in order to motivate recommendations to improve their utility to DoD. More detail is found in Appendix B.

Social network models represent relationships among individuals, the flow of information among individuals, and other aspects of the ways by which individuals are connected to and interact with each other. These models are based on graph theory, and because of that, traditional operations research flow and network models have been used for analysis. The nodes in such a model are individuals and the arcs are derived from relational data—that is, who knows whom, who works with whom, and so forth. There are three forms of analysis in this area: traditional social network analysis, link analysis, and dynamic network analysis. However, all of these forms of analysis, at their core, involve graph theoretic concepts and computations.

Traditional social network analysis mainly involves statistical analysis to identify the topology of the network, the influential nodes, and the key positions in the network. Link analysis is concerned with pattern recognition in the network and used to look at the formation of cliques and other relationship groups. Dynamic network analysis adds simulation to traditional social network analysis and link analysis to look at network evolution. These three techniques are used to analyze relational data—that is, data about whether entities of one type relate to entities of another type. Successful analysis is heavily dependent on the existence of reliable data and the availability of computational resources.

Social networks are a promising tool for studying many problems of importance to DoD, such as terrorist networks, the spread of infectious agents, flows of information and influence within enemy forces, and others. While considerable progress has been made in recent years, there are a number of weaknesses in the current state of technology for social behavior networks. A few of the most important limitations are these:

- Existing visualization techniques do not scale well, and interpretation of the results is dependent on the particular visual representation of the network.
- There is no agreed-on set of metrics for social behavioral networks, and those that do exist frequently do not correlate well with the property being measured.
- There are no standard techniques dealing with missing or erroneous data, and since these networks are data greedy, missing or erroneous data are a common problem.
- There is insufficient ability to link social networks to other events and locations, which is necessary to ensure that such networks are not used in isolation.

In contrast to social network analysis, multiagent systems can be used to model the way in which social behavior emerges from the actions of a set of agents. Multiagent systems are computer-based simulations of a set of actors, called agents, that take autonomous actions as they interact with one another. The agents can be heterogeneous; for instance, some may represent humans while others represent groups with which the humans interact. There are numerous such systems, which may differ in the number of agents, the type of algorithm used, the cognitive and social sophistication of the agents, and whether or not the agents are constrained to move on a grid.

Multiagent systems are also limited given today's technology:

- Validation is difficult (and is discussed separately in the subsection "Expanded Concepts of Validation").
- The goal, producing multiagent dynamic network systems tied to empirical data, now requires a multiperson, multiyear data collection effort on top of a similar development effort.

Recommendation 6: DoD should devote significant research to social behavioral networks and multiagent systems because both are promising approaches to the difficult modeling challenges it faces.

Serious Games

A video game is a mental contest, constrained by certain rules, played with a computer for amusement, recreation, or winning a stake. A serious game, by contrast, is a video game designed and used to further training, education, health, public policy, or strategic communication objectives. In addition to the story, art, and software aspects they share with video games, serious games are designed to educate, and they have a basis of pedagogical knowledge.

There are strong driving applications to which DoD could apply a science of games if such a science existed. In the health domain, medical personnel could train via games to perform procedures on vital systems. This capability is already evident in the game America's Army, which contains a full three-lecture series on lifesaving during combat. Training and simulation using game technology and creativity is an obvious option for the DoD; but to realize the full potential of this option, DoD should do better to coordinate its efforts. At present, game development for defense training is being handed off to individual contractors, who are not necessarily tightly coupled to DoD requirements. Those arm's-length contracted efforts lead to systems that are not well connected to the ever-changing requirements of the military. Additionally, the games are being built with proprietary technologies and resources that cannot be reapplied to other parts of DoD without additional payments.[15]

To avoid these limitations and to advance its capabilities, DoD might consider forming a university-affiliated research center (UARC) focused completely on game-based training and simulation. Such a UARC should be coupled to universities strong in human performance engineering and in game development. DoD could also license a commercial game engine for any DoD purpose or develop its own open source game engine. Art and other gaming resources built at that UARC should be easily shareable for all DoD game-based training requirements. With such a UARC in place, it would be feasible to produce game-based systems for mission rehearsal deployable by the soldier in minutes rather than months. In addition, the next-generation combat modeling and analysis systems might be accessible through gamelike interfaces rather than complicated menus and submenus. While the committee is impressed with the successes achieved by existing UARCs, it recognizes that despite the difficulties with commercial development of DoD training games noted above, other approaches are possible, including efforts based at private companies or other nonacademic organizations or by the establishment of a research consortium. The goal is to give DoD adequate assurance of top quality and modern technology with appropriate cost and continuity and without conflict of interest.

To realize the potential of games both serious and for entertainment, it will be necessary to undertake research and development aimed at transforming the production of games from a handcrafted, labor-intensive effort to an effort having shorter, more predictable timelines, with increased complexity and innovation in the produced games and a stronger focus on their pedagogical effectiveness. R&D is needed in a number of areas:

- *Infrastructure*. The underlying software and hardware for interactive games, including multiplayer game architectures, game engines, streaming media, next-generation consoles, and new wireless and mobile devices.
- *Cognition*. Theories and methods for the modeling and simulation of computer characters, story lines, and human emotions and of innovative play styles.
- *Immersion*. Technology for engaging the game player by means of sensory stimulation, including theories of presence and of sensing a player's physical state and emotions.
- *Serious games*. Game evaluation, human performance engineering, and principles common to the different domains in which games may be applied.

Recommendation 7: DoD should form a research center or consortium focused on game-based training and simulation.

A more complete discussion of serious games can be found in Appendix A.

[15]The America's Army game was built using a commercial game engine and cannot be repurposed to other DoD requirements without licensing, at high cost, the game engine. The Army allows it to be used for other defense purposes but assesses very high return on investment (ROI) charges.

Network Science

The nascent study of networks per se, which would allow understanding them intrinsically rather than through particular instantiations, also shows promise as a foundation for advanced defense MS&A. Society depends on a diversity of complex networks, and this report has emphasized the growing dependence of the military on networks for information dissemination, command and control, and effects-based operations, among others. Despite this dependency, our fundamental knowledge about networks is in its infancy; indeed, there is no body of knowledge that can be called "network science." A recent Army-sponsored report (NRC, 2006), referred to below as *Network Science,* is probably the first attempt to define both the need for and the substance of a science of networks. Although the report does not specify a rigid body of knowledge to be incorporated in the new field, it defines network science as consisting of the study of network representations of physical, biological, and social phenomena leading to predictive models of these phenomena.

Network Science identifies research areas of special interest to the Army that, in addition, apply more broadly to the entire DoD. One high-priority area is modeling, simulating, testing, and prototyping very large networks. Other aspects of networks that are relevant to MS&A include the impact of networked structures on organizational behavior (see the subsection "Social Behavioral Networks" in this chapter) and on enhanced networked-centric mission effectiveness (see the following subsection). In agreement with this report, *Network Science* concluded that advances in network science can address the threats of greatest importance to the nation's security.

Recommendation 8: DoD should support and extend initiatives to cooperate with other agencies funding research on networks.

Building the Scientific Base for Embedded MS&A

Present efforts to design and use complex, dynamic models are hindered by major gaps in the theoretical underpinnings of such models. For instance, the mathematical formalisms used in most modeling assume that the system being modeled is closed—that is, that the model output will always fall into a clearly defined space of possible outputs. But this assumption is violated for MS&A embedded in other systems, because the models must account for inputs from a dynamically changing set of sensors.

The NSF has identified a number of advances needed in mathematics and statistics as part of its Dynamic Data-Driven Applications Systems (DDDAS) program, mentioned earlier in this chapter. For example, the DDDAS characteristic of allowing new data to be incorporated into running algorithms raises fundamental questions about the stability of those algorithms and their outputs. NSF's program has developed fundamental analytical challenges for understanding and managing DDDASs:

- The creation of new mathematical algorithms with stable and robust convergence properties under perturbations induced by dynamic data inputs.
- Algorithmic stability under dynamic data injection/streaming.
- Algorithmic tolerance to data perturbations.
- Multiple scales and model reduction.
- Enhanced asynchronous algorithms with stable convergence properties.
- Stochastic algorithms with provable convergence properties under dynamic data inputs.
- Handling data uncertainty in decision-making/optimization algorithms, especially where decisions can adapt to unfolding scenarios (data paths).

Embedded MS&A systems must revisit a number of mathematical and statistical issues. These issues are given new prominence because the interaction between models and live data sources may cause small effects to cascade. These issues include

- Assessment and propagation of measurement error.
- Combining different types of uncertainties.
- Adapting to small sample sizes, incomplete data, and extreme events.
- Evaluation of quantization schemes.
- Optimization or satisficing within complex solution spaces.

These issues are well known in the optimization community and have been extensively studied, but embedded systems face the additional challenge of adapting to the rapid and unpredictable changes resulting from new data or a new base model of the solution space.

The use of MS&A in embedded situations creates a need for mathematical methods to enable the evaluation of partial or intermediate results. Embedded MS&A systems must be able to reason about how closely they have approached a good-enough solution in order to evaluate the trade-off between better results and the investment of additional sensing and computing resources. This requires measures of goodness that are meaningful in spite of uncertainty about the achievable end state, the means of dynamically adjusting the streams of input to move toward that end state, and the rates of convergence to some quality target. There might also be competing criteria of goodness. A step toward this capability would be the development of new analytic methods that could (1) characterize the solution space so that designers know something about its areas of sensitivity, boundaries, bad areas, and well-behaved areas and (2) characterize (using sensitivity analyses) the impacts of assumptions within the models or simulations that are running. Reasoning about the progress of a given intermediate solution may also be thought of as an online process, where one is asking whether either more data or more time will yield a better solution. New mathematical models of computational processes might be helpful for this.

In addition to the particular problems of embedded systems there are other areas where mathematical advances would find ready application:

- The use of structured random search and determination of problems amenable to such an approach.
- Scalability of mathematical methods, including better partitioning, abstraction, and aggregation methods.
- Methods for analyzing the results of model composition, interoperability, and resource integration, including methods for combining different formalisms, different definitions of uncertainty, and techniques such as metalogics, which allow one to reason about the characteristics of different logics and knowledge representations and their applicability to a specific problem.
- Methods for analyzing integrated modeling and data analysis environments. These would include research into the behavior of real-time linkage of models to data streams, perhaps using Bayesian methods to update model parameters with a combination of real and simulated data. Useful methods might also link machine-learning techniques for data extraction with simulation tools for forecasting.

This report has identified embedded MS&A as an important component of the changing DoD landscape and as an area in need of additional scientific research.

Recommendation 9: DoD should begin cooperative programs of research into embedded systems with other agencies facing similar demands.

Expanded Concepts of Validation

Some M&S is, or could be, solidly based in settled theory or empirical testing. Classical validation methods would then apply, and a model's predictions could be compared against a trusted reference. In addition, successful analysis requires confidence in the model's results, or at least an understanding of the limitations of the results. This said, many models and simulations, especially when considered to include the databases with which they will be used, contain a great deal of uncertainty (see Chapter 4 for an extensive discussion of this). If the data for previous events are known, at least retrospectively, then postdiction can be used to assess validity, but even that condition is often not met. The output of complex systems is influenced by one-time occurrences along the way that cannot be identified reliably even after the fact—when, for instance, a soldier is killed while on sentry duty or a surveillance aircraft is shot down without an impending or actual attack ever having been reported.

Even worse, when dealing with complex systems, we often do not even know what the correct structure of a good model should be. We may have reasonable conjectures, but it is hardly unusual for experts to disagree fiercely on such matters. For instance, traditional validation methods might not be applicable to large-scale multiagent models used for examining sociocultural systems because the fundamental underlying laws either do not exist or are unknown. Considerations such as these necessitate a new concept of validation;[16] it may be prudent to implement a means of labeling a model, simulation, or game as "valid for the purposes of exploration in a particular context." This would be a judgment not about the truth of any one prediction but about whether, on balance, the tool was useful.[17] Note that the important standard technique of judging face validity does not apply to the kinds of exploratory models and games on which this report focuses. Often the purpose of exploratory work is to uncover possibilities very different from what would usually be expected: system failures when certain odd combinations of events occur, such as long strings of (good or bad) luck; changes in the very structure of a social system due to personalities, deaths, random encounters at special times, and so on. If the model is used to find unexpected outputs, face validity is a poor judge of model validity.

How might one assess validity, even for limited purposes of exploration? The committee is skeptical about the value of bureaucratic processes to assess validity, since they are expensive, time-consuming, and frequently reinforce conventional wisdom and standard databases even when the reality is massive uncertainty.

Nonetheless, validity is an important matter. Several criteria are necessary to establish a model's validity:

- The model should be sufficiently consistent with the laws of physics and realities of technology so that the insights apparently obtained are not artifacts of violations of these.[18]
- The model (or game) should be comprehensible and explainable, often in a way conducive to explaining its workings with a credible and suitable "story," thereby helping people to assess in real time whether an insight could be illegitimate or the result of artifact.
- Models used in exploratory analysis should deal rea-

[16]Some M&S professionals feel that the term "evaluation" is preferable because it avoids the connotation, sometimes associated with "validation," of a one-time process or step of bureaucratic certification.

[17]Some of this discussion draws on Bigelow and Davis (2003).

[18]This is not an idle example. It is not uncommon for "concept-driven" war games to assume that technology will provide whatever is necessary to achieve the concept. That can be a useful approach, but it can also be troublesome, as when a concept dependent on light, long-lifetime, powerful chemical batteries is embraced uncritically.

sonably with all known classes of uncertainty, possibly deep uncertainty, and at least confront candidly the problem of unknown unknowns,[19] with some combination of speculation or stating of assumption.
- Models should be falsifiable. As in science, assertions and predictions that are ultimately circular should not be tolerated.

Multiresolution, multiperspective modeling can be very useful for validating troublesome models. One of the most convincing and economical ways to falsify some models is by looking at aggregate-level consequences and comparing them to aggregate-level empirical information. For example, if a detailed simulation shows complete military victory and successful stabilization with only a very small offensive force, then an excellent basis for skepticism is a low-resolution model reflecting historical experience that a much larger force had been deemed necessary in prior campaigns (Gordon and Trainor, 2006).

Another approach to such validation efforts is to work methodically through the components of a model, such as one used for exploration, examining the validity of the mathematics and logic in each component and the presence of factors known empirically to be significant.[20] Such testing is desirable when feasible, but the tester should be aware that because the modules of a complex system model are valid does not necessarily mean the overall system model is valid.

Even models of complex adaptive systems can be given the capability to explain results (e.g., by instrumenting the model or saving all relevant data so that a step-by-step replay is possible), although the current state of the art for doing so is poor. If such explanatory capabilities are built in, then conclusions can be evaluated in part by the chain of events leading to a particular result. Of course, a flaw in model logic does not necessarily indicate that the insight was wrong. Nonetheless, this method can be quite powerful when available.

Finally, sometimes a good way to assess model validity is to compare models and their "predictions" (even those of exploratory analysis) to the predictions of models built by other people, preferably with different mindsets. This is common in examining scientific disputes. The result may be to find important errors or omissions, to note significant differences without being able to evaluate relative correctness, or to find reasonable consistency—at least in a specific problem context.

[19]Deep uncertainty is sometimes said to be uncertainty of the type one has when even the nature of the underlying processes is unknown. A statistician might refer to not knowing the nature of the probability distribution. The Secretary of Defense has referred to "unknown unknowns," which are behaviors omitted from the modeling entirely because their existence is not recognized.

[20]This is not always straightforward. In social science it is not uncommon for some experts to insist that a factor is important, even though there is no empirical basis.

Games are even more problematic. Games are superb vehicles for revealing factors and considerations that might not otherwise be recognized and for building a "sense of the chessboard" and the moves that can be made. Some games also bring out a range of plausible and revealing human emotions, such as distrust and parochialism, and various misperceptions that are well-understood by cognitive psychologists. However, it seldom occurs to anyone who has played a game that the game should be validated. What would validation mean? Only one path through possibility space was traced out, and not everything happening in the game was necessarily realistic.

Nonetheless, games might provide a new opportunity for the validation of social behavioral models—especially if they include cross-cultural players, which present a particularly difficult problem for verification. We have no way to monitor humans and their behaviors such that those behaviors could be provided as inputs to a social model and could produce an output—that is, an action or behavior that a human or group of humans might perform. However, massively multiplayer online games (MMOGs) provide an environment in which experimentation and testing might be performed. By some estimates the number of players participating in online games already amounts to 180,000 person-years of game playing.[21] If these games could be instrumented and the behavior exhibited and captured, they could serve as virtual laboratories for the study of social phenomena. Recorded behaviors could then be tested against the outputs of social models.

The difficulty, of course, is that current MMOGs are commercial; their mission is to provide an entertaining and engaging experience to customers/players, not to run experiments of interest to DoD. However, it might be possible for DoD to carefully negotiate the funding of a virtual laboratory that would attach to a commercial MMOG and that could be used by DoD to test its behavioral models and by the game owner as an analysis tool.

The preceding discussion should be read not as suggesting a deemphasis on careful model evaluation but rather as urging recognition that "evaluation" must necessarily be quite different for models dealing with highly complex and uncertain phenomena than, for example, for engineering models. When dealing with issues that are less measurable but relevant to, say, effects-based operations, different methods are called for, and demands for validation in the classical sense are not pragmatic.

INFRASTRUCTURE TO SUPPORT THE NEEDED MS&A CAPABILITIES

The preceding section discussed capabilities needed for DoD's MS&A in order to address the challenges of Chapter

[21]Statistics available at http://www.gamasutra.com/gdc2005/features/20050309/postcard-diamante.htm.

2 and some promising technical directions to pursue. In practice, DoD's assessment of new MS&A capabilities will depend heavily on establishing a substantial forward-looking infrastructure. Laying the right infrastructure could have extraordinary benefits over the long run; failure to do so could greatly impede progress and efficiency.

To build the capabilities described so far in this chapter, the committee regards the following issues as the most important with respect to infrastructure:

- *Composability of M&S.* The ability to improve efficiency and coherence by constructing higher-level models from lower-level components.
- *Data collection and data farms.* The ability to draw quickly on existing databases, whether of input assumptions or previously generated output.
- *Visualization.* The ability to visually interpret high-dimensional data.
- *Chains of tools and computational platforms.* The ability to used linked chains of tools and platforms.
- *Service-oriented architectures.* The ability to develop and use modularized functionality that is available on the network as a service.
- *A definitive repository.* The existence of a central virtual repository and clearinghouse for pointers and advice.
- *Cooperation with other entities.* The ability to communicate across organizational and cultural boundaries unhindered by stovepiping or bureaucratic or technical obstacles.

This section explores these needs. Chapter 5 explores another important area of infrastructure, the educational background of the MS&A practitioners.

Composability

A recent technical review of model-composability issues (Davis and Anderson, 2004) discussed DoD model composability in some depth. The committee does not attempt to replicate the advice contained in that review except to highlight some of its main points and add commentary and recommendations. Appendix C provides more detail, including citations to the recent literature.

Composability is the capability to select and assemble components in various combinations to satisfy specific user requirements meaningfully. In M&S, the components in question are themselves models and simulations. Composability implies the ability to assemble components readily in various ways for different purposes. It goes further than interoperability, which may be achieved only for a particular configuration, perhaps in an awkward one-time lash-up. To put it differently, composability is associated with modular building blocks.

DoD's experience with composability has been disappointing, despite the considerable priority accorded it and the promise it showed. As discussed in Davis and Anderson, four factors affect model composability:

- Complexity of the system being modeled.
- Difficulty in defining when composite M&S will be used.
- Strength of the underlying science and technology.
- Human considerations, such as the quality of management, the existence of a community of interest, and the skill and knowledge of the workforce.

Davis and Anderson's review recommended a number of priorities and actions, which are summarized tersely in Table 3.3.

In addition to the priorities listed in the table, the committee offers the following general guidance on composability:

- To obtain the highest degree of composability, it must be engineered in. In general, DoD should treat composability as a matter of degree, measured as a function of the time and effort necessary and the flexibility obtained.
- By differentiating among (1) conceptual models, (2) implemented models, (3) simulators, and (4) experimental frames, the quality of MS&A can be substantially improved. There may be alternative ways to implement each of these, but without a clear distinction among them, it is hard to make sound judgments.
- DoD should continue to support the development of potentially standard ontologies, such as the Web Ontology Language (WOL), under development by the World Wide Web Consortium. These address key semantic and pragmatic issues important to the advancement of composability.
- Poorly documented legacy code will continue to be a challenge for composability into the indefinite future. DoD should invest in a selective program of retrodocumentation.
- In a few high-leverage cases, DoD should reprogram legacy models that appear to be valuable but that are technologically obsolete or limited.

Improved Data Collection for MS&A

Models of complex systems usually require large amounts of data as inputs, for determining parameters and for validation. Collecting those data can become a technological challenge. DoD needs to automate, or semiautomate, the collection of data for building and validating new models and simulation systems. Key tools for this more automated approach will be improved data-mining and text-mining techniques. Data-mining and text-mining tools are becoming in-

TABLE 3.3 Recommended Priorities for Improving Model Composability

Category	Component	Specific Priority Items
Science and technology	Military science for selected military domains	Capabilities-based planning, effects-based operations, network-centric operations
	Science and technology of M&S	Model abstraction (including multiresolution modeling model families)
		Validation
		Heterogeneous M&S
		Communication: documentation and new methods of transferring models
		Exploration mechanisms
		Intimate man-machine interactions
	Standards	Revisit standards, as in the pre-HLA days, but at the same time hurry to realign DoD's direction with that of the commercial marketplace
	Model representation, specification, and documentation	Exploit commercial developments, especially for high-level specification
Understanding		Develop methods to predict difficulty and cost of proposed composability projects
		Commission independent lessons-learned study on experiences from JSIMS, JWARS, and one SAF
Management		Define requirements and methods for developed first-rate M&S managers
Workforce		Stimulate systematic education, selection, and training of M&S workforce, in cooperation with other agencies, academia, and industry
General environment for DoD M&S		Improve incentives and mechanisms to improve industrial and other bases
		Encourage marketplace of ideas and assure even playing field for competitions
		As part of this, insist on transparency and exchange, reducing the scope of proprietary restrictions

SOURCE: Adapted from Davis and Anderson (2004).

creasingly important parts of the modern simulationist's tool kit. Such tools pave the way for more automated collection of the large time-sensitive data sets that are now needed and will be even more needed for simulation, particularly in the DIME/PMESII areas. To date, however, data-mining and text-mining tools are limited in the following ways:

- Lack of automated or semiautomated ontology creators to facilitate data sharing and analysis in new areas.
- Limited ability to handle nontext data such as photographs.
- Limited ability to extract data from various formats such as pdf and PowerPoint.
- Lack of good entity extraction algorithms that automatically correct for typos, spelling errors, aliases, and the like.
- Absence of tools that can work with streaming data.

For most MS&A of complex systems, there is a dearth of relevant data available in clean preprocessed form. Thus, to reduce the time spent by analysts on data collection and increase the time spent on analysis, automated and semiautomated tools for data gathering, cleaning, sharing, and other processing are needed. Such tools should include natural language-processing tools for extracting relational data from audio and text sources, Web-scraping tools, automatic ontology generators, and visual interpretation tools to extract network data from photographs and visual images. Appropriate subtools for node identification, entity extraction, and thesaurus creation are also needed. The development and availability of these tools in an interoperable environment is critical for providing masses of data that can be used for model tuning and validation. More rapid data collection would also mean the availability of more data sets for doing the meta-analyses required for development of theoretical foundations for M&S. Finally, these tools are essential to providing the wealth of data needed by models of complex, adaptive systems if models are to accurately represent situations and organizations and be the basis for sound analysis.

MS&A is turning to data-farming techniques to assess the response surface of a simulation.[22] Data farming is a technique where thousands (and potentially millions) of simulation runs are done with the same tool but using different parameters for each run. The "farm," or archive of data, is allowed to grow over time as the number of cases, even using related models and different data sources, grows. This results in a massive amount of data showing the behavior of a complex nonlinear model (the simulation) under a vast number of conditions. The output data are then automatically statistically analyzed and prepared for visualization, enabling the users to have a better understanding of system behavior. Insights are uncovered over time using sophisticated analysis tools and verified or disproved as the cases accumulate. This is critical for development and validation and for providing policy guidance. Such an approach is particularly valuable for system dynamic and multiagent MS&A efforts, where small changes in parameters can lead to large changes in outcomes.

Because many models of complex systems require such large amounts of data and long run times, the space of results cannot be adequately mapped using traditional experimental procedures. This may be true even after exploratory analysis, which carefully chooses scenarios to evaluate. By placing the models in a data-farming environment, the number of virtual experiments considered, the space of possibilities examined, and the scope of conditions analyzed can be expanded, often by several orders of magnitude, thus providing a stronger basis for decision making. Further, once a model has been evaluated as appropriate for the intended application, the response surface equivalent can be used, where appropriate, as a rapid model in training situations.

From the DoD perspective this approach is necessary if these models are to be used to provide actionable intelligence and support course-of-action (COA) analysis. While many models permit various COAs to be analyzed, the typical use of the models limits that analysis to between 10 and 30 or so different actions. To be sure, this is about an order of magnitude beyond what is done without the multiagent or system dynamic tools, where the number of COAs evaluated is typically fewer than 5.[23] However, with data farming, two or three more orders of magnitude might be achieved. The point is not so much the number of cases run but, rather, the ability to understand better the potential consequences of attempting a particular COA in different circumstances. For systems sensitive to many details, recognizing and characterizing the classes of circumstances requires extensive computation.

While the promise of data farming is great[24] and the number and diversity of MS&A systems that can be placed in a data-farming environment is increasing daily, most MS&A systems have not been placed in such environments, for four principal reasons:

- The cost of gaining access to and using such systems, particularly in terms of personnel training and machine computation time, tends to exceed the resources typically allocated for development and analysis.
- There is little guidance available on the extent to which data farming is needed and on how to make use of such technology.
- Placing a simulation tool into a data-farming environment often requires substantial code development in the simulation and sometimes in the data-farming technology; although plug-and-play is the vision, the overall technology is still very primitive.
- The amount of simulated data that can be produced in this way is much greater than what can generally be stored, handled with modern databases, analyzed with current statistical tool kits, or meaningfully visualized with most visualization systems. Current data-farming environments have been used only on relatively simple models to analyze only a few outcomes and have not been linked to databases.

Since data farming has such great potential for mapping the response surface of MS&A tools, the data-farming simulator must be able to reason about the trade-off between the precision of results and computation time and must implement those decision rules in the systems. Tools will be needed to reason about whether or not more time yields much better answers or enables a meaningfully wider sweep of the response surface. Tools will also be needed for automated data compression and visualization when the level of fidelity needed in the results requires a massive number of runs. (Note that these requirements overlap those identified earlier as needed for MS&A that will be incorporated in embedded systems.) Finally, one of the key limitations is computation speed. Multithreading, gridding, and parallelization all

---

[22]If there is a theoretical basis for postulating an approximate structure to be tested and calibrated statistically, then the result may be a "motivated metamodel," although in this case the purpose may be to summarize the implications of an ensemble of cases run with a relatively simple agent-based model rather than the behavior of a trusted detailed model.

[23]This technology is directly relevant to exploratory analysis as discussed in the subsection "Exploratory Analysis." As an example, suppose that a strategic/operational campaign model is to be used for exploratory analysis. If 10 parameters of the model are of particular consequence, then exploring the space of possible values with low, mid, and high values would mean 310 (more than 58,000) cases. Experimental-design techniques can reduce the cases, but the numbers are still very large, making the ability to use high-performance computing and data farms very attractive.

[24]See, for example, Barry and Koehler (2004) and Sanchez et al. (2004).

reduce computation time, as do special-purpose integrated circuits for common nonlinear analyses. However, such techniques are still very much under development and not yet ready for the typical DoD user or for potential developers in most universities or small businesses. Hence additional efforts are needed to automatically transform code into multithreaded, grid-based, or parallelized versions, and chips that support rapid nonlinear processing need to be routinely embedded in the laptops and desktops used by developers.

Visualization of High-Dimensional Data

In general, commercial off-the-shelf visualization techniques are not yet available for high-dimensional data and dynamic data. Specific short-term military needs include the following:

- Ability to receive visual alerts to significant changes in data streams.
- Ability to overlay or gracefully move between social network and geospatial information.
- Ability to zoom through large networks (on the order of one million nodes) and drill down on information about individual nodes.

Development of open-source visualization tools would speed the development and testing of other tools, because currently a great deal of effort is spent developing specialized tool kits for visualization. That is, most modelers have found that for their model to be accepted, they need good visualization. Consequently, the production of visualization tools for the interface sidetracks them from (and draws resources from) their primary goal of developing and testing new simulation tools. The result is a large number of suboptimal visualization tools designed for specific simulation engines. Common visualization tools are needed, just as we now have common statistical and database tools that can be employed by all simulation tools.

Chains of Tools and of Computational Platforms

To advance the functionality and applicability of multiagent systems (MASs) for defense purposes, the MAS and network analysis techniques must be integrated into tool chains—that is, linked analysis techniques that may include M&S along with other methods. An example of such a tool chain would be a pattern-discovery technique, used to derive equations from historical data, feeding into an MAS to evolve future systems. MAS techniques can be used to evaluate COAs and suggest areas for further data collection. Combining these techniques will enable new types of problems to be solved; for instance, combining social-network metrics with pattern-discovery techniques is key to building an understanding of how networks grow and evolve.

This is not to suggest that DoD should move to large, integrated behavioral models—quite the contrary. What is needed is increased interoperability of the tools. MAS development frameworks and the explosion of network analysis tools is making social behavioral modeling much more widely available. Moreover, it is leading to the development of many small single-purpose tools. DoD should be taking advantage of this by encouraging interoperability. It is important to note that it would not be feasible to require all tools to be written in a single language or to use a single framework; rather, the integration of models from diverse domains and in diverse languages is needed. Multiple models, visualization tools, and related software should be available to address diverse problems, but in a way that data (real and virtual) can be easily shared among the various tools.

There are a variety of things needed to support such interoperability. Standards for the interchange of relational data need to be developed. Behavioral modeling tools need to be Web-enabled, and XML-based input/output languages need to be developed. A uniform vocabulary for describing relational data needs to be developed; this is particularly critical as tools and metrics are coming out of at least 20 different scientific fields.[25] For defense and intelligence applications, common platforms and data-sharing standards should be developed so that tools written in the unclassified realm can be rapidly moved, without complete redesign, to the classified realm. To enable interoperability, a common platform and common ontologies for these tools are needed. This, in turn, will allow novel problems to be addressed more rapidly by regrouping existing models. It will enable subject-matter experts to interact through their models, thereby facilitating a broader approach to problems and reducing the likelihood of a biased solution. Interoperability will also hasten rapid development and deployment.

DoD operates in a dynamic and heterogeneous environment. Traditionally, the deployed forces have had minimal, if any, computational resources of note. However, as new systems are deployed, the warfighter will have access to computational resources at various levels. In some cases, such as some of the C4I and mission-planning devices, these are general-purpose computers. In others, they are embedded in the systems, which are moving to commodity processors. This is exemplified by the recent decision of the Future Combat System program to choose commodity processors rather than a traditional embedded processor. This shift to computational power available at all echelons and locations opens up new opportunities for the application of MS&A to support operational missions in real time.

The availability of computational assets, linked via communication networks, will allow for the implementation not

[25]These fields include anthropology, sociology, psychology, organization science, marketing, physics, electrical engineering, ecology, biology, bioinformatics, health services, forensics, artificial intelligence, robotics, computer science, mathematics, statistics, information systems, medicine, civil engineering, and communications.

only of the concepts of NCOs but also of distributed M&S. This can be accomplished through some combination of two methods. The first is the use of the current data in the operational systems as the starting points for the simulations. The second is the use of increased connectivity to implement an echelon-based model for running simulations and doing analysis. The latter option is discussed here in more detail.

While the concept of edge computing is not new, the ability of deployed forces to implement it is. In the edge-computing model, computational power is pushed to the most forward elements. At each level back to the institutional computational environments (most likely high-performance computing resources), there is a significant increase in the computational power available, as measured by both cycles and bandwidth. Thus, the ability to run large or computationally intensive simulations increases in proportion to the distance, in a network sense, from the front. Systems on the edge would tend to focus on computing time-critical information, while systems in the rear compute less time-critical but more computationally intensive elements of simulations. As an example, systems at the platoon level might compare ongoing actions to the plan, while systems in the rear are computing alternative COAs by doing simulations. When the operation deviates sufficiently from the original plan, the new plan has already been developed and is ready for dissemination.

In many ways, this paradigm emulates the real process of command and control by having the execution element at the front edge concerned with the immediate and the element to the rear concerned more with the longer term. Complex MS&A for theater use must be compatible with, and take advantage of, this distribution of computation along a chain of platforms.

If MS&A is to meet its potential, models of different but appropriate levels of fidelity must be operable at different levels of the chain of command and produce consistent results. In addition, these models must be effortlessly linked to other analytical tools and data sources, again appropriate to the level of the decision.

Service-Oriented Architectures

A key technology for addressing the interoperability and information sharing required by chains of tools and of platforms is the service-oriented architecture (SOA). In this context, the term "service-oriented" refers to customer service, not the branches of the armed services. An SOA mediates information exchange by means of services offered by service providers and used by service consumers (MacKenzie et al., 2005). Services are advertised and accessed by participants in a standardized way. Key issues for an SOA are visibility (the need for participants to see the resources offered by other participants), interaction (the processes by which those with needs and those with capabilities are matched with one another and can exchange information), and effect (the changes in the world that result from the interaction). A service is the mechanism by which needs are matched with capabilities and information exchange is effected. By itself, an SOA does not provide a solution to a domain problem. Rather, an SOA provides a means of organizing, composing, and delivering solutions and/or parts of solutions owned or controlled by various parties. Because the SOA concept is based on the market paradigm of autonomous agents exchanging items of value, it can be expected to scale more successfully than traditional top-down architecture concepts.

In all but the simplest situations, meeting a consumer's need typically requires invoking multiple capabilities offered by different sources and composing their outputs in some manner. To accomplish that, an SOA would have what is called a solution composition capability. A solution composition capability performs the following functions:

- Transform a need statement provided by the consumer into a problem description framed in terms of capabilities offered by service providers.
- Decompose the problem into subproblems that can be addressed by the available capabilities.
- Select appropriate capabilities to address the subproblems.
- Request services to exercise the needed capabilities.
- Receive outputs and combine them into a solution to the original need.
- Transform the solution into a form that can be accepted by the requestor.

A fundamental requirement for a successful SOA is semantic interoperability between providers and consumers. Semantic interoperability requires more than just the ability to interchange a given type of data in a given format. Provider and consumer must attach the same meaning to the data being exchanged.

Semantic interoperability is typically addressed through metadata and ontologies (Chandrasekaran et al., 1999). Metadata, or data about data, provide descriptive information about an entity of interest. While metadata have classically been used to represent structure within a single database, an SOA can use metadata to describe aspects of any resource, including access mechanisms, required policies, and provenance. Standardized metadata vocabularies with agreed-on semantics have evolved in many communities and can enable discovery and retrieval of relevant networked resources. An ontology is a formal representation of knowledge about a domain, typically expressed in a manner that can be processed by machines. Ontologies represent the types of entities that can exist in the domain, the properties these entities can have, the relationships they can have to one another, and the events and processes in which they can participate. However, current-generation ontology languages such as OWL have no means of representing uncertainty in ontologies. Because uncertainty is ubiquitous in DoD

MS&A, there is a need for ontology formalisms and languages capable of representing uncertainty associated with entities and their properties, interrelationships, and associated processes.[26]

Ontologies help to ensure that information is interpreted by consumers in the manner intended by the provider. Ontologies can also be used to reinterpret data. A simple example is unit conversion, as when a consumer specifies a need for an aircraft capable of transporting a given number of pounds and a provider specifies cargo capacity in kilograms. Ontologies would be referenced by the SOA to perform the appropriate compositions and transformations. Provider and consumer could each use their own units, and the service mediating the information exchange would be responsible for the conversion.

Ontologies and metadata alone cannot solve the semantic interoperability problem. First, legacy systems might have idiosyncratic and nonstandard interface specifications that bear no obvious relationship to current standard metadata vocabularies and ontologies. Developing wrappers to translate between representations is typically labor-intensive and error-prone, and a fully faithful translation may not be possible. Second, an SOA needs to mediate between diverse communities of users, each with different needs, different knowledge, and different customized vocabularies. Any attempt to force common vocabularies and ontological commitments on participants is doomed to failure just as surely as would be an attempt to force all the world's population to speak a single language. Third, even if it were possible to enforce a single common interface standard today, the standard would soon be out of date. New needs would inevitably arise that could not be met by the standard, and demand for change would escalate. Finally, populating metadata and developing ontologies is time-consuming and expensive. Thus, metadata and ontologies may be incomplete and out-of-date. The market paradigm underlying the SOA concept is instructive in this regard. Markets are ever changing, and successful products evolve with the changing market.

A Definitive Repository

The central repository for DoD's M&S community, the Modeling and Simulation Resource Repository (MSRR), is maintained by the Defense Modeling and Simulation Office (DMSO).[27] Rather than containing the actual data, it contains a series of user-entered records that describe each resource and where that resource is located. As the designated lead for DoD M&S, DMSO led the development of the MSRR as the clearinghouse for M&S resources across DoD. As a distributed system, there are 10 major nodes where the information is maintained. In addition to DMSO, each of the services has its own node,[28] as do the Missile Defense Agency[29] and the intelligence community.[30] Special interest groups such as the Joint C4ISR Decision Support Center,[31] the Object Model Resource Center,[32] and the Modeling and Simulation Information Analysis Center,[33] also have their nodes on the system. Collectively, the MSRR sites are supposed to encourage cost saving and cost avoidance by providing a framework within which DoD M&S activities can share resources. More of a card catalog than a repository, the nodes provide pointers to where the user can search and then follow up with the owners of the systems, data, and resources in order to obtain more information or the actual data.

As can be inferred from this description, the MSRR reflects, and is limited by, the structure by which DoD's M&S is managed. The real resources are distributed and available only to those who know their location and have the ability and personal relationships to acquire them. This is evidenced by the experience of one of the authors of this report. In order to acquire what amounted to an off-the-shelf database, it was necessary to make a personal call to a member of the engineering staff who had developed it (whose name was known only from viewing a demo), coordinating among three government agencies, and waiting three months for the paperwork to wend its way through the system.

Access to models is often the same, or worse. While some of the models are programs of record, are formally maintained, and have a baseline and a defined distribution mechanism, quite a few are not. These latter models are often maintained by contractors who view them as their own intellectual property and are accessed on the basis of individual task orders to support specific events. Thus, there is no real configuration control board or development plan for them. As a result, it is often difficult to know exactly what is in a model or how something was implemented. This makes software reuse a difficult task, as the model that is available might not be the most current or, alternatively, the changes made during reuse might not be rolled over into the next version. The result is a large number of local baselines for the more common models.

That there is no true centralized clearinghouse for M&S systems or data seriously hinders efficiency and reuse. With the problem compounded by many variations of models in common use, it is not surprising that M&S practitioners often feel it is easier to create or extend an in-house model or database than to reuse an existing one.

---

[26]See, for example, Costa et al. (2003).

[27]See http://www.msrr.dmso.mil.

[28]See http://afmsrr.afams.af.mil; http://www.msrr.army.mil; http://nmso.navy.mil.

[29]See http://bmdssc.jntf.osd.mil.

[30]See http://umsrr.dmso/mil.

[31]See http://extranet.itis.osd.mil/dsc/indes.shtml.

[32]See http://omrc.msiac.dmso.mil.

[33]See http://www.msiac.dmso.mil.

Cooperation with Other Entities

A final infrastructure issue is the collection of factors that limit cooperation between the DoD MS&A community and other federal agencies and nongovernmental institutions with relevant expertise. Of course, one of these factors is that DoD's mission is perceived to be unique. This is certainly true for the warfighting mission, but DoD's interest in DIME modeling and logistics modeling, for instance, has considerable overlap with interests found in the State Department and the Department of Homeland Security. Moreover, DoD could share its basic technologies and expertise with the many agencies that are seeking stronger capabilities in MS&A.

One policy issue that limits cooperation with universities, and hence DoD's ability to leverage more academic contributions to MS&A, is intellectual property (IP). For example, an examination of the University of California's IP policy[34] reveals that even if a company were to pay all direct and indirect costs of an R&D project (e.g., a project to develop new technology for M&S), it could at best obtain a royalty-bearing license for the life of any U.S. patent generated by the R&D, after paying an issue fee and a minimum annual amount. The same rules would apply for foreign patents, but only if the company agreed to reimburse the university for all costs involved in the patent filing and maintenance.

A more common position is exemplified by the IP policy of the University of Georgia,[35] which states that it is the university's policy to retain rights and requires a license to exploit any invention or technology commercially. University employees have to assign the rights to their IP-based work to the university if university resources were used in any way.

In industry, it is customary for prime contractors and subcontractors to jointly own the IP rights when they collaborate in technology development. On government contracts, the government is normally given "government rights," which allows the contractor to retain the right to commercially market the product while the government can use it for its own purposes. The notable exception to this is the Small Business Innovation Research and Small Business Technology Transfer programs, where the companies retain the IP that they develop.[36]

These disparities in IP policies are a problem because they discourage industrial partners from funding academic research rather than doing it in-house or with a commercial entity. If the project is expected to generate any significant revenue, the royalty to the university would reduce the organization's profit. Furthermore, the publish-or-perish mentality at most academic institutions will lead to disclosure of the innovation and could speed a competitor's time to market. Because the potential long-term economic disadvantages often argue persuasively against working with academics, DoD may be losing the talents of some of the nation's best scientists and engineers.

The issues of IP, publication, and subject matter are not the only issues limiting interactions between system developers and academics. With the large number of foreign students in technical fields, export control regulations are a significant concern. Given the subject matter of most military simulations, International Traffic in Arms Regulations (ITAR) restrictions basically prevent any non-U.S. citizen from having access to the models without export agreements in place. This is a further disincentive to collaboration between academia and industry on operational systems.

[34]See http://www.ucop.edu/raohome/cgmanual/chap11.html#11/340.

[35]See http://www.ovpr.uga.edu/rpph/rph_chp2.html#Ownership%20of%20Intellectual.

[36]See http://www.acq.osd.mil/sadbu/sbir.

REFERENCES

Barry, Philip, and Matthew Koehler. 2005. "Simulation in context: Using data farming for decision support." *Proceedings of the 2004 Winter Simulation Conference*, Washington, D.C.

Bazaraa, M.S., H.D. Sherali, and C.M. Shetty. 1993. *Nonlinear Programming: Theory and Algorithms*. New York, N.Y.: Wiley.

Bigelow, James, and Paul K. Davis. 2003. *Implications for Model Validation of Multiresolution Modeling*. Santa Monica, Calif.: RAND.

Brown, G.W. 1951. "Iterative solution of games by fictitious play." *Activity Analysis of Production and Allocation*. New York, N.Y.: Wiley.

Bui, H., S. Venkatesh, and D. Kieronska. 1998. "A framework for coordination and learning among teams of agents." *Lecture Notes in Computer Science* 1441:164-178.

Cares, J. 2006. *Distributed Networked Operations*. Newport, R.I.: Alidade Press.

Carley, K.M. 2003. "Dynamic network analysis." *Dynamic Social Network Modeling and Analysis: Workshop Summary and Papers*. Washington, D.C.: The National Academies Press.

Chandrasekaran, B., J.R. Josephson, and E. Benjamin. 1999. "What are ontologies and why do we need them?" *IEEE Intelligent Systems* 14(1).

Costa, P.C.G., K.B. Laskey, K.J. Laskey, and M. Pool. 2005. *Proceedings of the Workshop on Uncertainty Reasoning in the Semantic Web, International Semantic Web Conference*, November. Available at http://ftp.informatik.rwth-aachen.de/Publications/CEUR-WS-Vol-173.

Davis, Paul K. 2003. "Planning for adaptiveness." *New Challenges for Defense Planning: Rethinking How Much Is Enough*. Santa Monica, Calif.: RAND.

Davis, Paul K. 2006. "New paradigms and new challenges." In M.E. Kuhl, N.M. Steiger, F.B. Armstrong, and J.A. Joines, eds., *Proceedings of the 2005 Winter Simulation Conference*. New York, N.Y.: Association for Computing Machinery.

Davis, Paul K., and Robert Anderson. 2004. *Improving DoD's Model Composability*. Santa Monica, Calif.: RAND.

Davis, P.K., and J.H. Bigelow. 2003. *Implications for Model Validation of Multiresolution, Multiperspective Modeling (MRMPM) and Exploratory Analysis*. Santa Monica, Calif.: RAND.

Davis, Paul K., Jonathan Kulick, and Michael Egner. 2005. *Implications of Modern Decision Science for Military Decision Support Systems*. Santa Monica, Calif.: RAND.

Dupuy, Trevor. 1987. *Understanding War: A History of the Theory of Combat*. St. Paul, Minn.: Paragon House.

Garcia, A., D. Reaume, and R.L. Smith. 2000. "Fictitious play for finding system optimal routings in dynamic traffic networks." *Transportation Research B, Methods* 34(2).

Ghate, A.V., M.A. Epelman, and R.L. Smith. 2005. *Sampled Fictitious Play for Complex Systems Optimization,* Technical Report 05-15. Ann Arbor, Mich.: Department of Industrial and Operations Engineering, University of Michigan.

Gordon, Michael, and Bernard Trainor. 2006. *COBRA II: The Inside Story of the Invasion and Occupation of Iraq.* New York, N.Y.: Pantheon.

Holland, John. 1995. *Hidden Order: How Adaptation Builds Complexity.* Boston, Mass.: Addison-Wesley.

Hughes, Wayne. 1989. *Military Modeling.* 2nd ed. Alexandria, Va.: Military Operations Research Society.

Johnson, Stuart, Martin Libicki, and Greg Treverton, eds. 2003. *New Challenges, New Tools for Defense Decisionmaking.* Santa Monica, Calif.: RAND.

Lambert, Robert, Steven Popper, and Steven Bankes. 2003. *Shaping the Next One Hundred Years: New Methods of Quantitative, Long-Term Policy Analysis.* Santa Monica, Calif.: RAND.

Lambert, T.J., M.A. Epelman, and R.L. Smith. 2005. "A fictitious play approach to large scale optimization." *Operations Research* 53(3).

MacKenzie, C.M., K.J. Laskey, F. McCabe, P. Brown, and R. Metz. 2005. *Reference Model for Service Oriented Architectures.* Available at http://xml.coverpages.org/soa.html.

Military Operations Research Society (MORS). 2004. *Operations Analysis Support to Network Centric Operations.* Minisymposium held on January 27-29.

Nagel, K., M. Stretz, S. Leckey, R. Donnelly, and C.L. Barrett. 1997. *TRANSIMS Flow Characteristics*, LA-UR 973539. Los Alamos National Laboratory, Los Alamos, N.M.

Nagel, K., R.J. Beckman, and C.L. Barrett. 1999. *TRANSIMS for Urban Planning,* LA-UR 984389. Los Alamos National Laboratory, Los Alamos, N.M.

National Research Council (NRC). 2006. *Network Science.* Washington, D.C.: The National Academies Press.

Patek, S.S., V. Venkateswaran, and J. Liebeherr. 2001. "Simple alternate routing and differentiated services networks." *Computer Networks* 37(3-4):447-466.

Potts, R.B., and R.M. Oliver. 1972. *Flows in Transportation Networks.* New York, N.Y.: Academic Press.

Robinson, J. 1951. "An iterative method of solving a game." *Annals of Mathematics* 54:298-301.

Sanchez, S.M., T.W. Lucas, and T.M. Cioppa. 2004. "Military applications of agent-based simulations." *Proceedings of the Winter Simulation Conference*, Washington, D.C.

Sterman, John D. 2000. *Business Dynamics: System Thinking and Modeling for a Complex World.* New York, N.Y.: McGraw-Hill.

U.S. Government Accountability Office (GAO). 2006. *Defense Acquisitions: DoD Management Approach and Processes Not Well-Suited to Support Development of Global Information Grid.* GAO-06-211. Washington, D.C.

Waldrop, Mitchell. 1992. *Complexity: The Emerging Science at the Edge of Order and Chaos.* Simon and Schuster.

Weigley, Russell. 1973. *The American Way of War: A History of the United States Military Strategy and Policy.* New York, N.Y.: Macmillan.

4

MS&A and Decision Making

All MS&A activities within the defense establishment must eventually be responsive (and appropriately linked) to DoD decision makers and decision-making processes. In its deliberations, the committee considered how best to ensure the responsiveness of MS&A to the needs of those decision makers. The broader mission space now facing DoD implies that decision makers will rely increasingly on MS&A, and improving this interface is critical in order to profit from the recommendations of Chapter 3. In particular, this chapter discusses how to better match DoD's MS&A models and activities with the specific requirements of the problem; how to improve the interactions between the MS&A team and the decision makers; how to match MS&A activities and products with the styles of the decision makers; how to better quantify and manage uncertainties; and how to document and communicate the results of MS&A.

IDENTIFICATION OF THE DECISION PROBLEM AND SELECTION OF AN MS&A APPROACH

Defense decisions come in many varieties, including those that affect military strategy, technology acquisitions, and personnel management, as well as real-time decisions on the battlefield and its training equivalents. Clearly, different decision problems will require different MS&A approaches, so that selecting an approach that matches the needs of the problem to the needs of decision makers is essential.

Matching MS&A with the needs of decision makers should be a deliberate and interactive activity but not necessarily an extensive and time-consuming one. It should be taken seriously and involve direct interactions between the decision makers, the MS&A team, and eventual users of the MS&A products. According to vonWinterfeldt and Edwards (1986), three steps can be distinguished in this activity:

1. *Identifying the problem.* This step addresses a number of questions: What is the nature of the problem? Who is the decision maker? What decisions are to be made? What groups are affected by the decision? At this stage, simple lists of alternatives, objectives, events, and rough formal relations among them are created. To sharpen the sense of conclusions that might be salient and nontrivial, it is often useful to list potential conclusions and to imagine contradictory conclusions, so as to avoid biases and highlight difficult or controversial issues.
2. *Selecting an MS&A approach.* In this step, the problem identified in the preceding step needs to be matched with an MS&A approach—for example, simple or relatively complex modeling, deterministic or probabilistic simulation, one-sided analysis or a game-theoretic analysis, optimization or exploratory analysis, rational-analytic decision analysis versus subjective portfolio balancing, and fixed-model or system-dynamics model. To do this matching, the MS&A team should ask: What are the main problem complexities that the MS&A activities are intended to address? What is the purpose of the MS&A activity? Which MS&A approaches have been previously used successfully for this type of problem?
3. *Developing a detailed MS&A approach and architecture.* This step involves the more familiar territory of fleshing out the specific MS&A models, simulations, and analysis tools. Tools like influence diagrams, decision trees, and flow charts are used for this purpose, depending on the MS&A approach chosen. The committee suspects that the bulk of time, effort, and resources devoted to many, if not most, defense-related MS&A activities has traditionally focused on this last step, but it stresses that without proper attention paid to the first two steps, these resources might well be misplaced.

Although these steps have been developed and proven successful in the context of large, complex, and strategic decisions requiring analytic support, a streamlined version is likely to be useful even for short-term, quick-and-dirty activities as well as for preparing MS&A for training and exercise in tactical contexts.

These three steps may take a few days to several weeks, or longer, if trust must be gained from scratch. Previous successful uses of MS&A in similar problem contexts can shorten this time. For complex strategic decisions, for which no precedent exists, iteration between the MS&A team and the decision makers at each step can produce insights for restructuring and simplification.

There exists little research to guide MS&A practitioners in these steps. It is fair to say that the first two steps are more of an art than a science, while some research support has been developed for the third step in specific MS&A subdisciplines. For example, guidelines for the third step have been developed for studying causal dynamic systems (Sterman, 2005), objectives hierarchies (Keeney, 1992), decision trees and influence diagrams (Clemen, 1996), and Bayesian belief networks.

A related issue is the choice of the level of detail for an MS&A activity. Often, a little modeling and analysis goes a long way. Rapid prototyping, followed by restructuring, followed by more detailed modeling is usually a better strategy than investing large resources in a one-shot, large-scale MS&A activity. An arguably ideal MS&A process would involve an iteration of models that are initially too simple, to models that are too detailed and complex, back to simpler models that capture the essence of the complex models yet strip details from them that are unnecessary for the final use by decision makers. In such an iterative process, models can be simplified in many ways, including aggregation of variables, approximate computations, and omission of unimportant variables. Since these simplifications have implications for the uncertainty and accuracy of the output, the MS&A team must understand the implications and communicate them carefully to the decision makers and end users of their products. The involvement of decision makers and end users in these choices is extremely important if the MS&A development is to be matched to their needs. The committee believes that this practice is already followed by the best MS&A practitioners but must become more widespread, especially as decision makers become more dependent on MS&A to supplement their experience and intuition.

INTERACTION BETWEEN THE MS&A TEAM AND DECISION MAKERS

One way to improve Steps 1-3 is to iterate with the decision makers and other users and stakeholders of the MS&A activities and results. The discussions in Steps 1 and 2 (define the purpose, scope, and approach) are often inadequate because the actual decision makers are misunderstood or overinterpreted. Access to the decision makers may be very limited, and the intent of the decision maker may be filtered through layers of staff. In addition, the modeler needs to obtain the trust of the subject-matter experts, staff, and the decision makers. As the detailed modeling approach evolves (Step 3), iterations are still useful, but they may not require involving the highest level of decision makers.

Spetzler (2007) describes an explicit process of interacting with decision makers in the so-called snake diagram (see Figure 4.1).

Another form of intensive interaction is "decision conferencing." Decision conferences usually assemble the key decision makers and a facilitator with the intent to structure and solve a decision problem in real time—sometimes in 1 or 2 days. The process is as much about interaction and rules for interactions as it is about the formal models that are used to support decisions.

Being flexible and adaptive is very important. In one case where MS&A was applied to inform the selection of one of several technologies for producing tritium for nuclear weapons (von Winterfeldt and Schweitzer, 1998), the modeling approach was changed radically in midstream, from multiattribute evaluation to probabilistic simulation, because the decision makers had gained insights from the initial analysis and the information they needed had evolved. The issue in this application was whether to produce tritium in new nuclear reactors, accelerators, or existing commercial reactors. At the same time, a site for the facilities for these technologies was to be selected. The initial formulation of the problem was to evaluate five technologies and six sites based on 23 criteria using a multiattribute utility approach. A midanalysis briefing by the Assistant Secretary for Defense Programs made it clear that site selection was a minor issue but that uncertainties about production assurance (timeliness and capacity), as well as uncertainties about costs, were critical complexities that needed to be better understood. As a result, the analysis was turned into a risk analysis of the key production assurance and cost uncertainties surrounding the main technological alternatives.

NORMATIVE VS. DESCRIPTIVE MODELS OF DECISION MAKING

Traditional normative, or rational, models for decision making include the subjective expected utility model (SEU), Bayesian inference models, and multiattribute utility models (for introductions to these models, see Clemen, 1996; Hammond et al., 1999; von Winterfeldt and Edwards, 1986). While these models form the foundation for decision making under uncertainty and provide the links to other forms of MS&A, it has long been recognized that they are not descriptive of how people make judgments and decisions. Therefore, a question arises about how to implement rational models in the face of the possibility that decision makers' decision styles and natural ways of thinking about problems might not agree with the rationality assumptions. Traditional decision analysis approaches have also been challenged be-

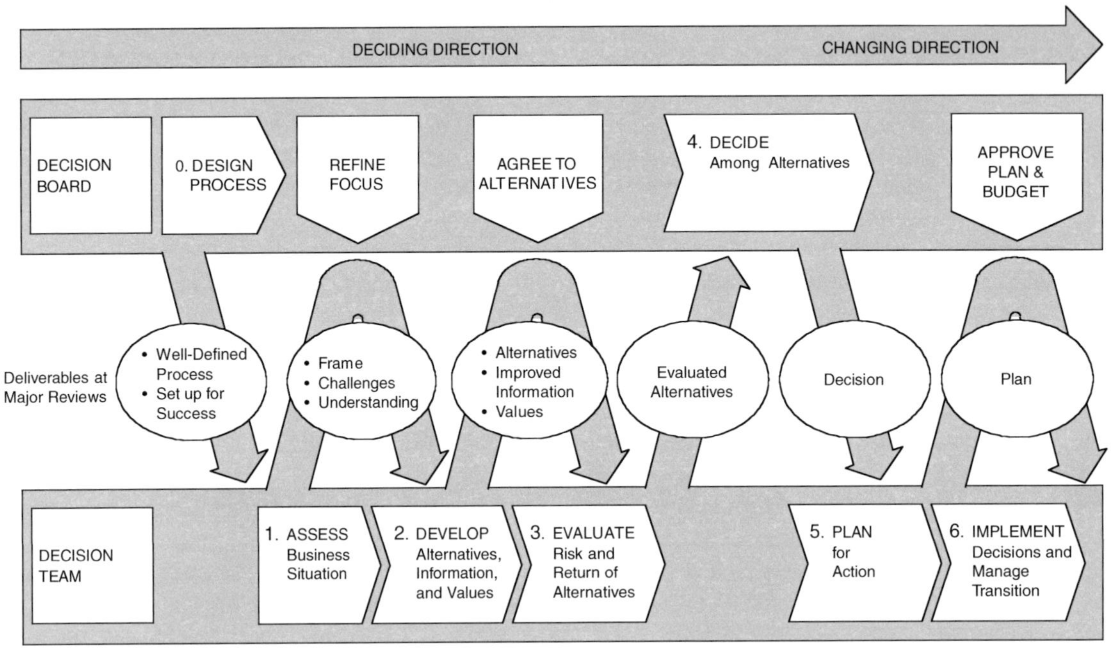

FIGURE 4.1 Snake diagram developed by the strategic decision group. SOURCE: Spetzler (2007).

cause of their limitations in dealing with problems characterized by deep uncertainty, which require strategies that are flexible, adaptive, and robust (Davis et al., 2005).

The literature on cognitive biases and heuristics, summarized in Kahneman et al. (1982) and Kahneman and Tversky (2000), is concerned with the dysfunctional nature of psychological aspects of judgments and decision making and how these biases and heuristics can lead people astray and prevent them from making sound judgments and good decisions. Subsequent research showed how simple heuristics and biases often can be very functional, approaching optimal analytical solutions to a surprising degree (Gigerenzer and Selten, 2002). An even more positive attitude is represented in the literature on naturalistic decision making (see Davis et al., 2005). Given the value of some of these naturalistic approaches to decision making, it is important to better understand the match, or lack thereof, between standard MS&A approaches and the naturalistic way decision makers think.

One suggestion is to combine analytic, rational approaches and intuitive, deliberative approaches to inform decision making, because neither MS&A alone nor unstructured deliberation without MS&A support is likely to succeed. Instead, an analytic-deliberative process that combines the strengths of both approaches may be the best way to find creative and acceptable solutions to complex decision problems.

Recommendation 10: DoD should strive to better understand the cognitive styles of decision makers and their interaction with different forms of MS&A. Research into decision-making styles would improve decision making with MS&A by affording intuition and insight into complex problems and enhancing the creativity employed in their solution.

ADDRESSING UNCERTAINTIES

Many MS&A models provide point estimates of forecasts, ordered evaluations of options, or optimal allocations of resources. Current MS&A practice often accompanies these deterministic results with sensitivity analyses to indicate (1) the robustness of solutions, (2) the sensitive parameters, and (3) the breakeven points. However, all MS&A modeling efforts face irreducible uncertainties, and these uncertainties must be characterized and made explicit in the course of the effort.

There are several kinds of uncertainties, each with its own complexities and challenges for characterization:

- Environmental uncertainty,
- Parameter uncertainty,
- Model uncertainty, and
- Deep uncertainty.

"Environmental uncertainty" refers to natural variations in the decision environment—for example, in weather conditions or in the random behavior of natural or engineered systems, such as earthquakes on a known fault line or failure rates of components of a weapons system. Often this type of uncertainty can be characterized using empirical data and frequency distributions. "Parameter uncertainty" refers to uncertainties about model parameters that are due to not knowing the precise value of these parameters. For example, the failure rate of a new component is often characterized by a parameter that is estimated from hundreds of previous failures of similar components. This parameter cannot be observed, but its probability distribution can be constructed or assumed using expert elicitation methods and empirical data. "Model uncertainty" refers to not knowing which model is most appropriate for a given phenomenon. For example, it is well known that the assumptions underlying an exponential model of component failure are violated both by system off-on cycling and by wearout. Quantifying such model uncertainties is a great challenge that has not been sufficiently well addressed by the MS&A community. "Deep uncertainty" refers to factors that are essentially not knowable currently and for which the relevant probability distributions are simply not known.

Characterizing, quantifying, and managing these uncertainties is a critical part of MS&A. Uncertainties are often hidden behind assumptions that are not spelled out explicitly. When uncertainties are explicitly addressed, they are often assessed by experts who are prone to overconfidence and other biases. Model and deep uncertainties are very difficult to assess and therefore are often ignored. The management of uncertainties through dynamic adjustments and adaptive response strategies is a topic of interest in many fields that is only now receiving significant attention by researchers.

The expanded mission space now facing DoD necessitates that the MS&A enterprise augment its skills for interfacing with decision makers. More or less traditional defense planning provides MS&A practitioners with good guidelines for presenting assumptions, indicating missions examined, displaying cost-benefit trades, and so on. But for nontraditional missions, the MS&A practitioner needs better tools and practices for presenting and explaining overlapping types of uncertainties, some of which can be very large. A body of experience has been built up in communities that evaluate environmental risks such as those posed by the siting of industrial plants, where the uncertainties are very large and technical assumptions are key yet such evaluations must be digested by decision makers who might misinterpret some types of probabilistic information. One such approach is to use a classification system to group elements relative to perceived uncertainty. The defense MS&A community can learn from individuals with expertise in risk communication. Whatever approach is taken, it is clear that the modeling and simulation of nontraditional missions will span a wider range of phenomena, including the full range of PMESII effects, than was customary in the past. It will be imperative that MS&A practitioners have a broader background and work more routinely on teams that cover the broader range of topics that must be modeled.

Defense modeling has always involved great uncertainty—the fog of war is a reality, not merely a convenient figure of speech. Modeling and simulation of nontraditional missions involves uncertainties of a different character than those associated with force-on-force modeling. Cultural differences lead to uncertainties about enemy intent; new autonomous weapons systems lead to uncertainties about the performance of our own weapons; gaps in intelligence lead to uncertainties about enemy capability. All of these need to be incorporated into DoD's modeling and simulation.

Recommendation 11: DoD should seek better methods to characterize, quantify, and manage the uncertainty inherent in all aspects of MS&A—including inputs, modeling assumptions, parameters, and options.

DOCUMENTATION AND COMMUNICATION

Documentation and communication are essential elements of good MS&A practice, yet there is little research supporting these activities. It is important that all elements of the MS&A activities be well documented, from the framework of the analysis, to the terms and assumptions used, to the results. The documentation should allow readers to trace every aspect of the models, simulations, or analyses, including sensitive and robust features and shortcomings and weaknesses. Simplifications should be highlighted. The importance of input parameters, if not explicitly analyzed through sensitivity or uncertainty analysis, should be discussed. The results should be provided at an aggregate level and at multiple levels of disaggregation.

Communication must take place at all levels in an organization, ranging from the immediate client for the MS&A activity to the ultimate decision maker. Often there are several layers between the immediate client and the decision maker. Sometimes there are institutions above the decision maker, like congressional oversight committees or courts, that can challenge and overturn the findings and recommendations of decisions derived from an MS&A study. It is important that the MS&A team be prepared to communicate at all levels and address the specific information need, possibly using different briefing materials.

The style of the experts or decision makers who receive the communication must also be taken into account. Some decision makers are willing to accept reasoned recommendations as long as they believe that the source is competent and trustworthy. Others want to challenge specific assumptions and numerical estimates. Yet others want to obtain a detailed justification and account of the complete analysis process, its assumptions, and results.

Most MS&A studies can be communicated at different levels of detail. At the highest level, the study simply communicates the conclusions (perhaps with a recommendation), with some minimal backup. A lower level might present an account of the pros and cons of all alternatives, which can often be presented qualitatively. Below this are many increasingly detailed levels of information. The MS&A team should be responsive to the decision maker's requirements at all levels of detail that are supported by the analysis.

Many decision makers do not like to be told what they should do; their job is to make decisions. The MS&A team's job is to present the information in a way that makes choices transparent and to clarify the crucial parameters on which the choice depends (decision makers typically consider many factors other than those considered by MS&A). Simple, transparent models that allow decision makers to control input parameters and immediately observe outputs can be an effective way to develop their confidence in models. If the underlying model is very complex and time consuming to run, simplified models that capture the essence of the complex underlying model should be used.

In short, MS&A practitioners should document their activities and results in a transparent and traceable way. At all levels of presentation, all terms and assumptions should be stated explicitly, sensitive and robust features should be identified, and shortcomings and weaknesses should be discussed openly. Strategies should be identified and implemented to increase the flexibility, adaptability, and robustness of MS&A activities and results. The MS&A community should work closely with communications experts to develop and institutionalize effective techniques to communicate MS&A results.

Decision making is a highly individual and often idiosyncratic process, so it should not be surprising that a disconnect often exists between the MS&A practitioner and the decision maker. Every effort should be made to ensure that both have the same understanding of the problem, the assumptions that are necessary to model and analyze it, the alternative solutions offered, and the uncertainties associated with each. This is an area that requires technical, verbal, and presentation skills, all of which need to be adequately represented on the MS&A team.

REFERENCES

Clemen, Robert T. 1996. *Making Hard Decisions: An Introduction to Decision Analysis*. 2nd ed. Belmont, Calif.: Duxbury Press.

Davis, Paul K., Jonathan Kulick, and Michael Egner. 2005. *Implications of Modern Decision Science for Military Decision Support*. Santa Monica, Calif.: RAND.

Gigerenzer, G., and R. Selten, eds. 2002. *Bounded Rationality: The Adaptive Toolbox*. Cambridge, Mass.: MIT Press.

Hammond, J.S., R.L. Keeney, and H. Raiffa. 1999. *Smart Choices: A Practical Guide to Making Better Decisions*. Cambridge, Mass.: Harvard Business School Press.

Kahneman, Daniel, and Amos Tversky. 2000. *Choices, Values, and Frames*. New York, N.Y., Cambridge, England: Cambridge University Press.

Kahneman, Daniel, Paul Slovic, and Amos Tversky. 1982. *Judgment Under Uncertainty: Heuristics and Biases*. New York, N.Y., Cambridge, England.: Cambridge University Press.

Keeney, Ralph. 1992. *Value-Focused Thinking*. Cambridge, Mass.: Harvard University Press.

Spetzler, C. 2007. "Building decision competency in organizations." *Advances in Decision Analysis: From Foundations to Applications*. W. Edwards, R.F. Miles, and D. von Winterfeldt, eds. Cambridge, England: Cambridge University Press, in press.

Sterman, John D. 2000. *Business Dynamics: System Thinking and Modeling for a Complex World*. New York, N.Y.: McGraw-Hill.

von Winterfeld, Detlof, and Ward Edwards. 1986. *Decision Analysis and Behavioral Risk*. Cambridge, England: Cambridge University Press.

von Winterfeldt, Detlof, with E. Schweitzer. 1998. "An assessment of tritium supply alternatives in support of the U.S. nuclear weapons stockpile." *Interfaces* 28.

5

Education, Training, and Professional Practice in Defense-Related MS&A

Education is central to developing and maintaining a high-quality professional MS&A community and is especially crucial as that community takes on the many challenges outlined in Chapters 3 and 4. Structural issues within the defense establishment affect DoD's ability to attract and retain well-educated modelers. This chapter assesses the academic preparation of the current MS&A workforce at DoD. In it, the committee makes recommendations for satisfying the education and training needs of the future workforce.

This report's primary emphasis is on competencies that professionals in various roles within the MS&A enterprise need to possess. Specific elements of a curriculum—courses, training modules—are not viewed as ends in themselves but as a means to develop key competencies. That is, a curriculum geared to a given professional role should give students the competencies they need for capable professional performance in the role and should be evaluated with respect to how well graduates demonstrate these competencies. In addition, MS&A educational programs used by DoD must provide the means for professionals to upgrade their skills as they move through their careers. The DoD human resources system, in turn, should be judged according to how well it matches individuals with positions utilizing their competencies.

The emphasis in this chapter is on the education and training of practitioners of MS&A, as distinct from the education of pure mathematicians or computer scientists. For this reason, a curriculum is judged successful to the extent that its graduates understand the role of MS&A within the overall DoD decision-making process and can play their respective roles within this process. Practitioners should be educated to understand that models and simulations must be designed, built, and used in a manner that provides timely and effective answers to questions of importance to the decision makers.

The ever-increasing complexity of the problems confronted by defense decision makers, together with the increasing ability of technology to solve them, makes it impossible to segment today's MS&A world into specialized roles played by different individuals. The roles played by MS&A practitioners within the MS&A life cycle were discussed in Chapter 1. In many cases, a single individual will play multiple roles; in other cases, different individuals playing different roles work together to support a given decision-making objective. The various roles played by MS&A practitioners were identified in Chapter 1 and are expanded on here:

- *Analyst*. An analyst formulates models and interprets them. He or she creates a formal representation of a real-world problem in a form that is amenable to computation. In addition, the analyst often serves as the domain expert on the modeling team. In this role, the analyst understands which factors are important and must be included in the model or simulation, the level of fidelity needed, and the types of displays that might be useful. After the simulation is run, the analyst will transform its output into conclusions that inform a decision, assess the strengths of the result and the uncertainties in it, and determine the importance of the assumptions incorporated in the model to the validity of the conclusions. Finally, the analyst is frequently called upon to explain the results to a nonspecialist, often in nontechnical language. Because of the variety of skills needed for this explanatory role, it is often filled by more than one person on the MS&A team.
- *Modeler/programmer*. A modeler/programmer transforms the formal representation created by the analyst into executable form. He or she ensures that the designs are modular, computationally efficient, and well documented and that configuration management is adequate and has well-defined subsystem interfaces. The increasing complexity of systems and the growing drive for reusability and interoperability have given rise to highly specialized implementation environments, so that the modeler and the programmer may be different individuals.
- *Implementers*. An implementer adapts the program for

execution, reviews and amends the experimental plan, runs the M&S through the cases of the experimental plan, observes and iterates, and executes the postprocessing needed to visualize the model outputs and produce comprehensible and usable results for the analyst.

- *M&S manager.* A manager hires, fires, and manages personnel; suggests or approves choices of M&S and related purchases, education, and training; draws upon outside consultants for checks of quality; and interacts with peers in meetings of M&S governance. A manager may practice some quality control in specific analysis projects. Some M&S managers may also have analytical depth and may then function more like the analyst-in-charge.
- *Consumer.* The consumer employs M&S to support military decisions. Some consumers may be military or civilian decision makers and/or managers with little technical training and only a rudimentary understanding of the internal workings of the models and simulations that support their decisions. In other cases, they may be equally competent in one of the other roles or may even have played one or more of the other roles at some point in their careers. In still other cases, they play all the roles, as when a decision maker builds a simple model on his or her desktop computer to support a key decision.

Different roles may require different skill sets, acquired through education or experience. For example, analysts and modelers require different skill sets—mathematics, operations research, and statistics for analysts, computer programming for modelers—which are typically taught in different educational disciplines. Consumers may require only a passing understanding of the details of any of the other roles but need a solid appreciation of the limitations of the MS&A activity, the assumptions made, and the conditions within which results hold.

In addition to depending on the ability of individuals to play their respective roles, the effectiveness of MS&A depends on their ability to work together toward a common objective. However, in practice, individuals in different roles sometimes do not mesh as well as they should, and the quality of decisions can suffer as a consequence. The best educational programs consciously address this issue. The problems that remain are due to a number of factors, including inadequacies in some education and training programs or in the organizational climate and deficiencies in the preparation of some practitioners. The committee agrees that in many cases these concerns are well founded and that well-designed education and training can contribute to a solution.

The committee has identified several competencies that it believes have not been given sufficient emphasis and that should be included in the education of MS&A practitioners in the defense establishment:

- *Documentation.* MS&A professionals need to be capable of developing, or guiding the development of, clear and useful documentation for their models, algorithms, and analyses.
- *Meaningfulness.* MS&A professionals need to be able to describe the meaningfulness of their model outputs. For example, is the output shown one among many possibilities? Is it an expected value of a family of outputs? What statistical properties does it have?
- *Audit.* MS&A professionals need to be prepared to subject their work to detailed audit and criticism and to use the results of such audits constructively.
- *Comparison with reality.* MS&A professionals need to understand the real world being modeled in sufficient detail to compare their results against the results that might be expected and to justify any divergence.

In the sections that follow, the committee addresses the question of how to ensure that education and training prepare people to work effectively with other people playing roles that are different from their own, as well as to recognize the limits of their own competence and to compensate appropriately by drawing on the expertise of people trained for different roles. The talents required of MS&A practitioners and the roles they play in the process change over the course of an MS&A study. These roles and the extensive training associated with imparting the competencies needed for each of them make the effective employment of the MS&A workforce a challenging task and require that DoD use its MS&A workforce optimally throughout the life cyle of each project.

DoD is clearly not the only organization that is dependent on MS&A, nor is it the only one that is concerned with best practices in building and using computational models. A good recent example of a community's standards is found in Jakeman et al. (2006), in this case the environmental community. Another example, focusing on community standards that are desirable in the area of regulatory analysis, is OMB Circular A-4 (OMB, 2003). In fewer than 50 pages, it lays out standards that all agencies should observe in conducting and presenting regulatory analyses. A similar document promulgated by DoD would be useful by imposing uniform standards of practice for MS&A.

A SURVEY OF TODAY'S EDUCATION AND TRAINING LANDSCAPE FOR MS&A

Here the committee surveys the vast array of programs having some relevance to defense MS&A. This section provides a broad overview of the different roles for which education and training are needed and of the many educational options available to MS&A professionals at various points in their careers. It also considers how those programs mesh with current career progressions and identifies the essential ingredients of a successful MS&A education.

Individuals working in MS&A have undergraduate degrees in a wide variety of disciplines, including computer science, computational science, operations research, statistics, systems engineering, industrial engineering, and other traditional programs in engineering, mathematics, or science. For managers of programs or projects with a significant MS&A component, education in engineering management is highly desirable. Education in human factors and the psychological and social sciences is important, especially for those involved in human-computer interface design, human-in-the-loop simulation, and the modeling of human behavior. Very few people come out of an undergraduate education with broad enough and deep enough knowledge to effectively pursue a career in MS&A. Generally, a B.S. allows entry into an MS&A career only as a technician. Suggested requirements for a master's-level program are given in the next section, "MS&A Curriculum." However, with today's emphasis on lifelong learning and the rapid evolution of M&S technologies, it is expected that all MS&A professionals will upgrade and expand their skills as their careers progress.

Through its internal system of academies and colleges, DoD offers baccalaureate and postgraduate education programs for uniformed personnel aspiring to careers as MS&A professionals. Undergraduate education programs available at the U.S. Military Academy at West Point, the United States Naval Academy, and the United States Air Force Academy include MS&A. Postgraduate programs are available at a number of military academies, including the Air Force Institute of Technology, the Naval War College, the Naval Postgraduate School, and the National Defense University. As an example, the U.S. Army Command and General Staff College has institutionalized simulation into the curriculum for all of its schools. As the college shifts its curriculum toward execution-centric learning, simulations are increasingly being used to expose future general staff officers to real-world scenarios that are relevant to current operations.

Over the past 20 years, a small number of civilian academic programs have emerged that specialize in defense-related MS&A topics. Most have been at the graduate level, although an undergraduate program is offered at Arizona Polytechnic, and a few associate degree programs prepare students to become simulator technicians. Furthermore, many programs in other disciplines such as computer science or operations research have concentrations targeted at individuals entering or working in defense-related MS&A. Thus, a wide variety of options is available to individuals initiating or continuing their education for careers in defense-related MS&A.

An important emerging technology is distance education. With the newer technologies for distance education, including streaming audio and video interaction (e.g., Pullen, 2000), students can interact synchronously or asynchronously with instructors and other students, gaining an experience that is close to classroom education. For mature students in some subject areas, distance education now appears to be at little disadvantage compared with traditional classroom instruction and has the advantage of convenience. However, some of the most critical skills for an MS&A professional include the ability to work in teams and the ability to interact effectively with consumers. To acquire such skills, some degree of in-person interaction is essential. Nevertheless, technology for virtual meetings, asynchronous discussion, and file sharing can, if used appropriately, greatly enhance the effectiveness of teams and can augment traditional avenues for interacting with model consumers. Both the technology base and the experience base for distance education are advancing at a rapid pace, although the educational community still has much to learn about the strengths and weaknesses of distance learning for different types of students and different subject areas.

A vital factor in the health of the MS&A profession is the infusion of new talent. The MS&A community within the defense establishment is affected by two trends in the field at large that are troubling in this regard. First is the perceived decreasing interest in mathematics and science education on the part of native-born American citizens. Second is the difficulty, especially since September 2001, for foreign nationals to obtain visas to study in the United States and to remain here after completing their education. Together, these trends raise concerns about the influx of new young MS&A professionals into the workforce.

MS&A CURRICULUM

The committee examined educational programs offered by a number of institutions of the military establishment, held discussions with MS&A practitioners and consumers, and consulted the M&S Body of Knowledge published by the DMSO Education Consortium in April 2004.

These sources reveal a common core set of competencies that should be possessed by professionals in the field of MS&A. The committee concludes that a master's-level education for an MS&A specialist should provide three kinds of competency: background, core, and specialized defense knowledge.

Background Competency

An MS&A professional should have fundamental competency in undergraduate-level mathematics, science, and computing:

- Mathematical topics include algebra, trigonometry, engineering- or physics-oriented calculus, linear algebra, and differential equations.
- Statistical topics include calculus-based probability, inference, and data analysis.
- Computing background includes basic computer lit-

eracy (basic competency in word processing and spreadsheets), ability to write programs in a computer language such as C++ or Java, and experience with at least one model-development and/or analysis environment such as Matlab, Mathematica, or a spreadsheet macro language.
- Science should include college-level biology, chemistry, and/or calculus-based physics.

These competencies are acquired by most students who complete an undergraduate degree in mathematics, physical science, or engineering. Those with other educational backgrounds should make up the missing elements before embarking on specialized training in MS&A.

Core Competency

Building on the background competency, an educational program for a well-educated MS&A specialist would provide a common set of core competencies:

- *MS&A life cycle*. Students should appreciate the role of MS&A in decision making, be exposed to the history of MS&A, and gain an understanding of different types of modeling methods. Students should be introduced to the life cycle of an MS&A study and gain practical experience with each of the steps in the life cycle. Through the use of both positive and negative exemplars, they should understand the value of an organized, systematic approach to carrying out the phases of the life cycle. Their education should provide practical exercises in which they perform all the activities inherent in the life cycle of problem formulation, conceptual model development, selection of software, simulation design, analysis plan, execution, analysis, presentation of results, and evaluation.
- *Deterministic modeling and optimization*. Despite the stochastic and complex nature of many military situations, many others lend themselves to deterministic models that are capable of being optimized or nearly optimized. In addition, such models are useful as approximations to more complex real-world situations, submodels of larger simulations, quick-turn models to meet short deadlines, or models of situations with small variance where deterministic techniques may be sufficient.
- *Continuous and discrete event simulation*. An important core competency is the ability to construct and execute simulations of both deterministic and stochastic discrete event and continuous systems. The typical undergraduate or graduate degree program would include a course in which students learn to develop a model, develop a computer representation for their model, design experiments to be conducted on the model, implement the model in a computer simulation environment, conduct experiments, and analyze and interpret the results.
- *Probability and statistics*. The changing mission space places greater emphasis on performing effectively in situations characterized by large uncertainties. Thus, a strong grounding in the fundamentals of probability and statistics is essential, as is the ability to apply statistical thinking to military problems. The core background described above includes basic competence in probability and statistics. MS&A practitioners must also be proficient in more advanced areas of statistics such as regression, analysis of variance, experimental design, and data analysis, including data combination, and have some exposure to stochastic processes (e.g., Markov processes). An understanding of the concepts of state spaces, dependency conditions, and elementary principles for dealing with risk and uncertainty is important, as is a knowledge of decision analysis. These topics will not be a part of the undergraduate education of many students but are essential ingredients in the toolbox of an MS&A professional.
- *Topics in computing*. The core background described above includes basic computer programming skills. Additional necessary capabilities include an introduction to data structures, software engineering, basic numerical methods, and analysis of computational complexity.
- *MS&A evaluation*. Evaluating how well a model or simulation serves the purpose for which it is designed is one of the most important activities within the MS&A life cycle. The MS&A community places strong emphasis on validation, verification, and accreditation. While there is good reason for a strong emphasis on effective evaluation, excessive concern with formal certification can detract from the overall purpose of supporting the decision maker. MS&A education should cover VV&A thoroughly and should stress its importance not as an end in itself but as a means for ensuring that models and simulations are properly evaluated and that there are mechanisms to guarantee that the most appropriate methods are being applied to a given problem. MS&A education should also prepare students for dealing with the reality of models that cannot be classically validated (see "Expanded Concepts of Validation" in Chapter 3).
- *Human-simulation interaction*. A simulation is useful only to the degree that humans can interact effectively with it. An MS&A professional should have a basic understanding of perception, cognition, and the interaction of humans with computers. As computing technology advances, a greater variety of interaction modalities will become important. As the state of the art in human-in-the-loop interactive simulations advances, an understanding of human-computer interac-

tion becomes more important. The ability to communicate the results of MS&A to decision makers in ways that highlight its relevance to military decisions is an essential skill. The ability to communicate uncertainty is especially important, not only including error bounds due to well-understood stochastic factors but also model uncertainty and deep uncertainty.

- *Modeling humans*. MS&A has traditionally focused on modeling physical systems or systems in which social and cultural factors can be ignored. Sometimes there are strong pressures to focus resources on modeling well-understood aspects of a problem, leaving other aspects unmodeled. As the importance of including social and cultural factors in models and simulations grows, it will become necessary to provide students with the tools for incorporating these factors when traditional MS&A is inadequate. The appropriate modeling technology differs in important ways from traditional M&S technology. Students should be exposed to the literature on effective technologies for exploratory modeling (e.g., Bankes, 1993) and for modeling human and social systems.
- *Managing MS&A*. Individuals managing programs and/or projects in MS&A require management education. While a course of study in business administration provides useful education for managers, management of highly technical programs requires knowledge that goes beyond what is taught in most business administration programs. Some degree of technical training is necessary for individuals managing highly technical programs. In addition, MS&A in DoD has some unique problems not likely to be addressed in civilian schools. One is the joint management of civilian personnel and military personnel, subject to frequent rotations among the latter. A second is the necessity to break down the modeling stovepipes, discussed in Chapter 2, that are artifacts of previous military requirements.

Defense-Specific Competency

The above list of core competencies provides a strong grounding in MS&A that is not specific to military systems. Many civilian programs whose target population is students aiming for careers in civilian MS&A cover the bulk of this core. At DoD, practitioners also need background in the kinds of problems to which MS&A is applied in that establishment—for example, training, acquisition, test and evaluation, and combat. Also specific to DoD would be a familiarity with the main DoD simulation tools.

Many students at the master's level, at least in DoD-sponsored schools such as the Air Force Institute of Technology or the Naval Postgraduate School, might wish to specialize in MS&A applied to particular areas, such as combat, training, or logistics. In addition to competence in the particular subject matter and in MS&A techniques appropriate to that subject matter, they should have at least a rudimentary knowledge of the DoD organization, including the military components, and Joint Force organizations, and the overall planning, budgeting, and acquisition processes.

The above discussions stress the importance of different roles in the overall MS&A process. As was pointed out in Chapter 4, requirements for nontraditional missions require that DoD draw on the body of experience from the larger MS&A community. It has already been pointed out in Chapter 3 that the ability to work on the full range of PMESII phenomena requires building the skills needed to work in cross-disciplinary teams.

As mentioned earlier in this chapter, formulators, implementers, and analysts require stronger mathematical and computer competency than consumers, but even consumers must have some degree of competence if they are to understand the value and the limitations of the results and to apply them effectively to decisions. Conversely, formulators, builders, and analysts must have some degree of competence in the problem domain if they are to avoid inadvertently and inappropriately altering the consumer's problem in the way they construct the model or analyze the data.

Recommendation 12: DoD must give its MS&A practitioners some exposure to all the topics in the core curriculum: the MS&A life cycle; continuous and discrete simulation; probability and statistics; topics in computing; deterministic modeling and optimization; MS&A evaluation; human-simulation interaction; modeling humans; and managing MS&A. Familiarity with these topics is essential for all practitioners, although the depth of knowledge needed will depend on a practitioner's particular role.

FOSTERING A STRONG AND EFFECTIVE MS&A COMMUNITY AT THE DEPARTMENT OF DEFENSE

Key to the effective employment of MS&A to support decision making at DoD is an organizational climate that fosters high-quality MS&A. There are many MS&A professionals within DoD, but distributed widely among many offices and programs and subject to different missions, cultures, and priorities. There would be value in DoD working to encourage a more integrated MS&A community. In this section, the committee suggests ways to foster and maintain such an organizational climate.

The educational programs examined by the committee stress the importance of effectively integrating different activities:

- Modeling (creating computable and/or manipulable representations—physical, mathematical, logical—of the real world).
- Simulating (implementing models that describe system behaviors over time, usually in computer code).

- Interfacing with the user, trainee, or analyst.
- Analyzing (making sense of the output of the simulation; translating outputs into statements about the real world that are relevant to the decision maker's problem).
- Evaluating (assessing the fit between the model and the real-world problem; identifying assumptions on which results depend; qualifying conclusions appropriately to account for problematic assumptions and/or inaccuracies).

There is a general agreement, both in academia and in the applications community, about the importance of an education that effectively integrates these activities. Segregating them and treating them as isolated components can result in neglect of important aspects of the problem and can seriously degrade the effectiveness of analyses in support of decisions. In practice, however, these activities are sometimes segregated into different bureaucratic units in an organization. Institutional barriers between units can interfere with the effective use of MS&A to support decisions.

Recommendation 13: DoD should ensure that its educational programs provide experiences in which students integrate the activities of modeling, simulation, analysis, evaluation, and communication to address real-world problems of importance to the consumers of the information.

Education is a career-long process for today's professionals. As in all fields of science, change is rapid in computing and information technology. With this in mind, the committee believes that mechanisms need to be in place for MS&A practitioners to maintain currency. Civilian MS&A practitioners clearly need continuing education to maintain their technical competence. This need is even more critical to uniformed MS&A practitioners who routinely serve in non-MS&A positions. Because these uniformed MS&A practitioners may have periods during which they do not practice their MS&A skills, they may lose some of their technical skills and be unaware of advances in their associated technologies. The Army, for instance, has two MS&A-related specialty fields: FA57 (simulations) and FA49 (operations research and systems analysis); but officers also need to serve in other assignments that may not involve MS&A in order to advance their careers. This practice of rotating out of MS&A assignments, while potentially disruptive, is actually critically important for keeping the MS&A community connected to the operating forces. With this in mind, the committee stresses the importance of having institutional mechanisms in place to refresh and upgrade the competencies of uniformed MS&A professionals who rotate in and out of MS&A positions. It is essential that MS&A professionals upgrade their skills to cover areas in which their initial education was less than thorough and to keep up with changes in the field.

Maintaining professional currency can involve many other activities in addition to or in place of traditional classroom instruction. Practicing MS&A professionals need to devote a certain portion of their time to reading current literature, attending conferences and lectures, becoming involved in professional societies, and other activities that provide regular interaction with other professionals and exposure to new ideas. It is especially important that they be exposed to uses of MS&A outside their own specialty. Some of the most important advances come from the transfer of technology from one field to another. An awareness of how MS&A is applied in other disciplines can be an important source of new ideas for defense-related MS&A. In this regard, a vibrant professional community with opportunities for exchange of ideas is essential to maintaining the health of the profession. The Military Operations Research Society (MORS) provides many national and local conferences, symposia, and workshops, as well as a journal and newsletters. The Society for Computer Simulation publishes the peer-reviewed archival journal *Journal of Defense Modeling and Simulation*. Nondefense MS&A professional societies also offer opportunities for the exchange of ideas, and many of them have special-interest groups devoted to defense-related topics, such as the Military Applications Section (MAS) of the Institute for Operations Research and Management Sciences (INFORMS). Unlike MORS, MAS is open to individuals without a SECRET security clearance.

Building and maintaining an effective MS&A community requires fostering communication and collaboration among MS&A practitioners in government, industry, and academia. As noted in the final subsection of Chapter 3, collaboration is impeded by export controls and security regulations. Controls on the flow of information are, of course, necessary to prevent sensitive information from falling into the hands of those who would use it to harm the United States and its interests in the world. Nevertheless, building a strong, effective, and self-critical MS&A community requires drawing on the best available talent. There are actions DoD can take to access this talent that would not compromise national security. For example, DoD could require developers to build unclassified versions of models and unclassified databases, or to segregate classified parts of models in separate modules, with unclassified alternative modules that can be substituted when the model is being used in a nonsecure environment. This practice could actually make an MS&A capability more secure by restricting the use of classified information to situations in which it is necessary to the modeling objective. Furthermore, bringing in the most qualified individuals to perform evaluations enables more thorough and capable validation of models and gives DoD the benefit of the strongest available expertise.

An important issue for DoD is the place of both traditional and continuing MS&A education in the career progression of military officers. As officers with education in MS&A progress up the command chain, they tend to move from doing MS&A themselves to leading and managing others who do it. Officers with a solid education and practical experience in applying MS&A to real-world problems move into the ranks of leadership with a good understanding of how MS&A can be most effectively employed to support end users. It is essential that obtaining this education and practical experience not come at the cost of eventual career advancement. Furthermore, educational opportunities need to be provided in a timely manner to support an officer's current assignment and to anticipate future career directions.

REFERENCES

Bankes, S. 1993. "Exploratory modeling for policy analysis." *Operations Research* 41(3):435-449.

Executive Office of the President, Office of Management and Budget (OMB). 2003. "Regulatory analysis." Circular A-4, September 17. Available at http://www.whitehouse.gov/omb/circulars/a004/a-4.pdf.

Jakeman, A.J., R.A. Letcher, and J.P. Norton. 2006. "Ten iterative steps in development and evaluation of environmental models." *Environmental Modeling & Software* 21:602-614.

Pullen, J.M. 2000. "The Internet-based lecture: Converging teaching and technology." *ACM Special Interest Group on Computer Science Education (SIGCSE) Bulletin* 32(3):101-104.

6

Moving Forward

This report lays out important improvements to practice, research areas, and educational requirements needed to address emerging MS&A and military challenges. Developing an in-depth understanding of the challenges and the solutions requires updating and strengthening the base of military science, in addition to the more straightforward refining of technology and tools.

DEPARTMENT OF DEFENSE'S NEED TO LEAD IN MS&A AND MILITARY SCIENCE

In Chapters 2 and 3, the committee recommends substantial effort and investment to build the base of M&S science to deal effectively with networking; embedded systems; the implications of interconnected complex, dynamic, and adaptive systems; and other emerging issues. The committee identifies numerous topics and approaches that appear to it to be important for this enterprise. It will be essential, the committee believes, for DoD to approach this research, and subsequent applications, in a scientifically and analytically sound way. Doing so will require something that does not yet exist—one or more organizations in DoD (including the uniformed services) that have a specific charter to encourage, nurture, and exploit the results of the research and changes that the committee recommends. These organizations need to be staffed with individuals having a high degree of technical competence, an understanding of military issues, and a passion for organizing and codifying knowledge (or helping to stimulate such actions).

DoD will need to nurture and sustain such fundamental advances, and it should identify which organization(s) should have this responsibility as part of their charter. As with much research, such investments should be seen as "common goods" justifying centralized support, whether by a service, a joint organization, or the Office of the Secretary of Defense. Although the Defense Modeling and Simulation Office (DMSO) has produced and supported a great deal of infrastructure development—for example, that of the High-Level Architecture (HLA)—it has not regarded investment in the science of MS&A as part of its charter. Either this should change or some other organization should be tasked with such investment and nurturing.

Research in MS&A is needed to improve the modeling of real combat, and the organization(s) charged with leading this effort should go one step further and lead the development of basic military science. By this is meant the study of the technical, psychological, practical, and other phenomena that constitute warfare and armed conflict.

Military operations today comprise a large number of interconnected and counterposing systems and variables that, in addition to characterizing kinetic and other direct modes of confrontation, include special operations, combat service support, logistics, medical support, intelligence, counterintelligence, psychological operations, etc. Noncombatant entities typically outnumber combat entities, and more now needs to be learned about noncombatant activity than traditional direct force-on-force activities.

Such a new science of military operations would by necessity incorporate the reality of the civilian presence in many forms, from informants during combat operations to insurgents during reconstruction phases. This makes it difficult to represent phenomenology and, most contentiously, to report outcomes or sequelae of engagement with even the most sophisticated of current MS&A capabilities. Moreover, it is reasonable to argue that activity taking place in the so-called criterion environment itself—i.e., the battlespace—is so complicated that it is an inherently unreliable target for validating combat models. The systems being modeled may be highly nonstationary as well as stochastic. If that is so, then the usefulness of conventional stochastic models could be severely limited. This so-called "validity problem" represents a formidable—even (some would argue) overwhelming—problem for today's military science base.

Such deterrents to representing the most straightforward, contemporary military problems mean that, arguably, the military science base is exploited today well beyond what can be justified. Moreover, as the battlespace continues to increase in complexity, and as outcomes rely less on mere attritional or maneuver success and more on multivariable interactions, new computational approaches will be needed to support the more complex MS&A required. This difficulty is exacerbated by MS&A's inability to represent network-centric phenomena.

Combat is viewed by many today as systems of interconnected systems or, more formally, as complex adaptive systems (CAS). These CAS generate properties that are not unlike those that have been examined over the past decade or so by scientists who are concerned with complex phenomena in economics, meteorology, and ecosystems and who claim to have the computational capability to do so adequately. Salient among these complex, adaptive properties are self-organization and emergence.

On the battlefield, "exploitable complexity" is the military's attempt to produce self-organizing and emergent effects through indirect (often very indirect) means. In fact, the generation of effects-based operations (EBO) is viewed in military science as having the potential to be important in the future, for reasons that are beyond the scope of this report. Nevertheless, in the future MS&A must be able to facilitate the discovery, derivation, and analysis of tactics, operations, and strategies that produce "intended emergence," which is precisely what EBO attempts to do. There are arguably few problems that are more important, or more difficult, to represent than those associated with EBO. To provide the necessary analytic support for EBO, it is essential to significantly accelerate the rate of progress in the entire science base underlying military MS&A, including both MS&A science and military science.

A program of research for military ends, including both MS&A science and basic military science, is unlikely to happen without DoD sponsorship and support. The area is not particularly attractive to outsiders and, even if it were, they would find it difficult to get access to the needed information. Defense MS&A is crucially important, requiring DoD support in order to avoid losing many of the benefits from DoD's efforts in the more generic aspects of modeling and simulation.

Recommendation 14: DoD should identify (or create) and charge an organization with responsibility for developing and supporting a program of research and development directed at improving and updating the base of military science for combat and noncombat modeling. That same organization would be responsible for effecting the recommendations on education that are called for in Chapter 5.

To be sure, there are many shortcomings even in the scientific base for traditional MS&A, as well as continuing debates about such things as the appropriate form of attrition equations, movement-rate equations, and so on. The committee does not mean to discourage continuing research in these areas but wishes to encourage the new MS&A challenges where no satisfactory base currently exists.

STEPS FOR ADVANCING MS&A IN ENGINEERING

As noted in Chapter 1, the NSF recently published a report of two workshops on simulation-based engineering science that examined the role of MS&A for U.S. science and engineering, including medicine, materials, and other scientific disciplines. While that report touched on defense applications, it did not bring out the three central themes of the current report—network-centricity, complex adaptive systems, and embedded systems.

The NSF report did, however, identify many of the same issues facing engineering MS&A as does this report, and there are some similarities in the recommendations as well. In particular, the NSF workshops and this committee have some similar conclusions:

- *NSF.* "Formidable obstacles remain in linking highly disparate length and time scales. . . ." The committee expressed DoD's necessity for overcoming these obstacles in Chapter 3 in the subsection "Multiresolution Modeling and Families of Models and Games" and proposed solutions in the section "Composability."
- *NSF.* "Verification, validation, and uncertainty quantification are challenging and necessary research areas that must be actively pursued." The committee comes to the same conclusion for DoD, with a different emphasis, in "Expanded Concepts of Validation" in Chapter 3 and "Addressing Uncertainties" in Chapter 4.
- *NSF.* "Research is needed to effectively use and integrate data-intensive computing systems, ubiquitous sensors and high-resolution detectors, imaging devices, and other data-gathering storage and distribution devices, and to develop methodologies and theoretical frameworks for their integration into simulation systems." The subsection "Improved Data Collection for MS&A" in Chapter 3 of this report examines this issue in detail, and the section "Building the Scientific Base for Embedded MS&A," also in Chapter 3, discusses the integration of real-time data into embedded systems.
- *NSF.* "Computer visualization will be integral to our ability to interpret and utilize large data sets. . . ." This agrees with the committee's finding, expressed in "Visualization of High-Dimensional Data" in Chapter 3.
- *NSF.* "Meaningful advances in simulation based engineering science will require dramatic changes in science and engineering education." The committee discusses this topic at length in Chapter 5, with a focus on changes needed in the defense environment.

The similarities between the NSF's assessment of MS&A

and the committee's own assessment reinforces the committee's Recommendations 4, 8, and 9—that DoD undertake joint research programs with other government agencies where appropriate, citing network science and embedded systems as two areas where such joint research would be possible.

Despite the similarities, neither of the two reports subsumes the other, but they mutually support the conclusion that MS&A is a vital tool for achieving military superiority and economic competitiveness. Where the substance of the two reports does intersect, there is agreement on the details of how to go forward.

INTERNATIONAL ASPECTS OF MS&A

The committee did not address coalition MS&A in any depth; however, a few observations can be made. Most of the standard MS&A used for training and analysis by both the United States and its coalition partners focuses on the kinetic aspects of warfare. However, other nations are now seriously engaged in pursuing the issues highlighted in this report:

- *Complexity.* The Technical Cooperation Program (TTCP), which links work of Australia, Canada, New Zealand, the United Kingdom, and the United States, has hosted symposia on agent-based modeling and related issues. Much of the innovative research highlighted at these symposia has been done in nations other than the United States. The work of Project Albert, initiated by the U.S. Marine Corps, has also had substantial international participation.[1]
- *Command, control, and networking.* Important work on the transformational aspects of command and control has been pursued in the NATO context, where, as would be expected, much of the emphasis is on interoperability and on identifying core principles and methods that can be the basis of coalition operations. A recent good example of this is SAS 050, *Exploring New Command and Control Concepts and Capabilities*, Final Report, prepared for NATO, January 2006, available on the Web site of DoD's Command and Control Research Program, http://www.dodccrp. org/SAS/SAS-050%20Final%20 Report.pdf.
- *Joint Forces Command.* Under its manager for the training of joint and coalition forces and interoperability, the Joint Forces Command has an aggressive program of engaging partners in model-driven exercises. In many cases, U.S. models are used exclusively and U.S. systems extended to represent foreign systems. However, some countries, notably NATO members with a long history of interoperability with the United States, have models of their own. Like the U.S. system, these tend to focus on the kinetic aspects of warfare.
- *Allied Command Transformation (ACT).* This is a subcommand of NATO; more information on it is available at http://www.act.nato.int/. Representatives of the committee visited its Future Capabilities, Research, and Technology (FCRT) subdivision in Norfolk, Virginia. They learned about the M&S activity in progress at the FCRT's Modeling and Simulation Coordination Section. That section supports ACT in developing and refining concepts, defining capability requirements, and devising experimental and testing programs. Its basic approach is to use commercial off-the-shelf software for concept development and exploration, which is less expensive (in both money and time) than using large, complex existing packages or building new custom software. Some of the commercial packages in use at the time of the visit included Extend, AnyLogic, @Risk, and StatFit. The committee thought that this approach, which is standard in industry but which it had not seen used elsewhere in DoD, had the potential for reducing the time and cost of simulation studies in many areas, thereby increasing the effectiveness of simulation as a management tool.

Some of the themes emphasized in this report, in particular Recommendation 1, which called for developing flexible, adaptive, and robust systems, will facilitate coalition MS&A work. This includes MS&A directly related to operations such as mission rehearsal and command and control. In particular, flexibility, adaptability, and robustness are needed for networking, composability, revised approaches to VV&A and, most generally, for extending MS&A to account for DIME/PMESII issues, many of which depend strongly on alliance or coalitional activities. Further attention to the future of MS&A should, if at all possible, devote considerable effort to international issues.

CONCLUSIONS

This report has noted that DoD's MS&A enterprise is widely distributed across many offices and programs with differing missions and priorities. Although there is value in diversity, there is also value in coordination, direction, and guidance. For that reason, an R&D office, as suggested in Recommendation 14, could benefit all of DoD with the most advanced thinking and insights in this complex area. This would be a natural adjunct to the coordinating role played by the Modeling and Simulation Coordination Office (formally DMSO), and it would create an intellectual common ground for the military MS&A community that would help it address the recommendations in Chapters 3 and 5. Recommendation 14 is addressed to high-level DoD planners.

[1]See http://www.projectalbert.org.

The committee has concentrated on areas that it regards as underdeveloped but vital for the future of DoD MS&A and in which there is promise of disproportionately large advances. This does not, however, eliminate the need for continued improvement in more traditional areas such as force-on-force modeling, logistics modeling, and transportation modeling. Recommendations 1 through 5 and Recommendation 11 highlight the directions that the committee believes will be most important for the new challenges facing the United States: FAR methods and better methods for modeling network-centricity, embedded systems, and the incorporation of uncertainty. Recommendations 6 through 10 recognize five of the research areas discussed in Chapters 3 and 4 as paramount: social behavioral networks, game-based training and simulation, network science, embedded systems, and cognitive decision making. In Recommendations 8 and 9, the committee suggests that research in network science and embedded networks be undertaken jointly with other agencies. If resources for research are constrained, these five areas should have the highest priority. These first 11 recommendations should be the responsibility of those individuals and organizations that manage DoD's MS&A research and development.

An educated and effective MS&A workforce is essential for both progress and practice, and in Recommendations 12 and 13 the committee gives guidance for their development. These two recommendations should be the responsibility of the human resource managers in the DoD MS&A community. They also give guidance to individual practitioners as they plan their own careers.

The committee believes that the future of MS&A depends upon progress in both infrastructure and research. It has attempted to give guidance in both areas, tempered by its view of future challenges and likely scientific advances.

Appendixes

A

Serious Games and Their Role in Defense Modeling, Simulation, and Analysis

Michael Zyda

DEPARTMENT OF DEFENSE MS&A GAME APPLICATIONS

There are strong driving applications to which the Department of Defense (DoD) could apply a science of games if one existed. In the health domain, we easily envision medical training on vital systems in game form. An example of this is in America's Army—the full three-lecture series on combat life saving is in the game, including the test you take in the real Army! When the player passes that test, he then can act as a combat medic in the game.

In the public policy domain, we foresee games similar to SimCity, maybe SimNavy, where resource allocation and policy change can be explored for their effect before implementation. This makes the very large assumption that the resource models underneath the game are accurate and verified.

The value of games for strategic communication has already been demonstrated by the America's Army game. With wartime recruitment at an all-time low for the Services, it is clear that more work in this domain is essential to reach eligible youth.

Training and simulation using game technology and creativity is an obvious direction for the DoD; but for this approach to become fully effective, it needs to coordinate its efforts more. Right now, game development for defense training is being handed off to individual contractors, who are not necessarily tightly coupled to DoD requirements. Such arm's-length contracted efforts lead to systems not well connected to the ever-changing requirements of the department. Additionally, those developed games are being built with company-proprietary technologies and resources that cannot even be repurposed for other parts of the department without exorbitant payments.

With a university-affiliated research center (UARC) in place for this technology, we would see game-based rapid mission rehearsal systems, games deployable by the soldier in minutes rather than months. We will see our next generation combat modeling and analysis systems with gamelike interfaces rather than complicated menus and submenus. We can imagine a training system that recreates a virtual Fallujah, with all the excitement and stress of the real historical battle, followed by a careful and considered game-based take on nation building. We can imagine building in situ resource utilization games that allow us to explore colonizing the Moon in order to control space for defense purposes. There are clearly many potentials for a UARC with game focus.

With the development of the Army's SIMNET system, starting in 1983, the era of modern modeling and simulation began. The signature of the new era was the inclusion of the three-dimensional (3-D) visual display in subsequent modeling and simulation systems. Mostly gone was the era of building large computational models whose outputs were printouts delivered to analysts for pronouncement of results. 3-D visual displays had become mainstream for the DoD modeling and simulation (M&S) world by 1990.

Beginning in 1997 with the publication of the National Research Council report *Modeling and Simulation—Linking Entertainment and Defense*, it became clear to DoD that the entertainment community, in particular the videogame community, was generating better-performing visual systems than defense contractors (Zyda and Sheehan, 1997). The entertainment industry was producing highly immersive games and location-based entertainment, with wonderful visual displays, great performing artificial intelligence (AI) characters, and networking scales equal in size to those required by defense. The delivery of this report was a shock to the DoD in that historically defense had been the technology leader but now the entertainment world might be overrunning that position.

By 1999, DoD moved into high gear to attempt to catch up or at least to get access to this new force for its M&S requirements. In early 1999, the chief scientist of the U.S. Army asked the chair of the committee that wrote the report just mentioned to draft an operating plan and research agenda for an organization that was to become the Institute for Creative Technologies at the University of Southern California (USC ICT). The mission for USC ICT was to focus on defense and entertainment immersive technologies for use in Army training. One of the many projects USC ICT began was the development of the Full Spectrum Warrior and Full Spectrum Command videogames, delivered in the summer of 2003.

At the same time and in parallel with the formation of USC ICT, the U.S. Navy formed the MOVES Institute, whose mission was research, application, and education on the grand challenges of modeling, virtual environments and simulation. The MOVES Institute carries out research on 3D visual simulation, networked virtual environments, computer-generated autonomy, human performance engineering, and game-based simulation. The MOVES Institute developed its highly successful game, America's Army, and posted it on the Internet on July 4, 2002. America's Army became the fastest growing online game of all time and started a much larger discussion on the use of game technology and creativity inside DoD. Everyone wants their next-generation training system to be as beautiful and as easy to operate as America's Army. America's Army became the first delivered serious game to have a major impact, and its continued impact crosses beyond the boundaries of defense to the corporate world, where interest in serious game production is also large. There is great potential for transforming the military's MS&A efforts with serious games if that technology and creativity is deployed appropriately and a framework, or science of games, is created to support deployment. Serious games are not going to achieve their potential if we fall into the same patterns of hype and magic as prevailed in the early days of artificial intelligence and virtual reality. Before we can begin to define a research agenda for the science of games, we need a few definitions.

WHAT IS A GAME, AND WHAT IS A SERIOUS GAME?

The word "game" is emotionally charged, with the strength of the emotion breaking along the generation gap issue: Did you play videogames growing up? We define a videogame as a mental contest according to certain rules, played with a computer, for amusement, recreation, or winning a stake. We define a serious game as a mental contest according to certain rules, played with a computer, which uses entertainment to further the objectives of government or corporate training in fields such as education, health, public policy, and strategic communication.

A typical organization for developing videogames for entertainment and serious purpose is illuminating. Bing Gordon, chief creative officer at Electronic Arts, thinks of games as "story, art, and software." We learn that there is a design team, headed by a lead designer, who is responsible for the story, the entertainment component of the game. Then there is an art team, headed by the lead artist, responsible for the look and feel of the game. Finally, there is a programming team, headed by a lead programmer, responsible for developing code that implements story requirements, interface features, networking, Web connectivity, scoring systems, AI scripting, game engine changes—just about anything technical and programmatic required for the entire development effort (Zyda et al., 2005). Note that serious games have more than just story, art, and software. Serious games have pedagogy too—the activity of educating or instructing or teaching—activities that impart knowledge.

The activities of educating or instructing or teaching that impart knowledge or skill is exactly what is added to games that makes them serious. Now, notice that pedagogy has to be subordinate to story. Story is the entertainment part and that comes first; once that is worked out, then we can do the pedagogy. Pedagogy insertion, as it's called, comes from a human performance engineering team that works closely with the design team. There is a lead for that team, the lead pedagogist, who is a combination of instructional scientist and subject matter expert for the domain for which we are building the serious game. We cannot just build serious games by tossing their development to a traditional game team. That team has to interact with the instructional scientists and subject matter experts who make up a larger human performance engineering team.

Clearly, a research agenda that supports serious games also supports the entertainment industry, one of the largest industries in this country. In fact, the serious games research agenda is larger than the research agenda of the entertainment industry in that it has to carefully deal with the issue of merging pedagogy and story in videogame form.

CREATING A SCIENCE OF GAMES

The development and wide release of the America's Army game began a revolution in thinking about the potential role of videogames in nonentertainment domains and started a discussion on how to advance the state of the art of game technology to support entertainment and serious games of the future (MOVES Institute, 2004). DoD's application domain interests for serious games include games for modeling, simulation, and analysis, games for training, and games for strategic communication. To carry out that widespread deployment of games, we need to define a research agenda that will get us to the science of games.

A GAMES RESEARCH AGENDA

To impact the future of serious and entertainment games, we need to undertake an R&D agenda that transforms the game production process from a handcrafted, labor-intensive effort into an effort having shorter, more predictable production timelines, increased complexity, and innovation in the produced games. We see several components of that research agenda:

- Infrastructure,
- Cognition and games,
- Immersion, and
- Serious games.

Infrastructure

Infrastructure is the underlying software and hardware necessary for the development of the future of interactive games. Infrastructure includes work on

- Massively multiplayer online game architectures,
- Game engines and tools,
- Streaming media,
- Next generation consoles, and
- Wireless and mobile devices.

Architectures for massively multiplayer online games (MMOGs) are important for many application domains, including the military, homeland defense, and online education. The fundamental research question is how do we develop software architectures that are dynamically extensible and semantically interoperable. That is, how do we build game or simulation clients that can connect into a running MMOG, download the appropriate code for display and interaction, and then operate with the other online players? This is a question of interest to the gaming world and the large government gamc-based simulation world. There are currently no dynamic solutions to this—only static solutions that dramatically drive up the cost of large-scale simulation and gaming. We need to solve the MMOG architecture problem not just for game clients but also for large-scale computational architectures such as grid computing.

Game engines and tools are an important research area if we are going to attack the problem of lack of reuse in gaming and if we are to move games from crafted systems built by game industry technicians to engineered systems used widely in the government and corporate worlds. Currently the only part of the game world that uses reusable game engines is "first-person shooters." Some attempts to broaden that usage to other domains have occurred in the America's Army project, but they have all such suffered from major limitations (Zyda et al., 2005). Those limitations include the lack of support for large terrain boxes (many game engines can only handle 1 km × 1 km, and most real-world applications require much larger spaces), onerous and expensive game engine licenses, and the general lack of game engines for the R&D and serious games community at large. There is a need for an open source game engine, including a development tool set that is widely available and utilized, such as Linux. With an open source game engine, we can explore additional capabilities not provided now, including the larger terrain box, dynamic terrain, physical modeling, and other requirements ignored by the entertainment world. In addition, with an open source engine and testbed, other inadequately explored directions such as the modeling and simulation of computer characters, story, and human emotion, become possible.

Cognition and Games

We use cognition and games to develop theories and methods for modeling and simulating computer characters and story, modeling and simulating human emotion, analyzing large-scale game play, innovating new game genres and play styles, and integrating pedagogy with story in the interactive medium of games. Work in cognition and games includes

- Computer-generated autonomy,
- Modeling and simulating human emotion,
- Understanding and analysis,
- Pedagogy/story integration, and
- Game play innovation.

Computer-generated autonomy is the modeling of human and organizational behavior in networked games. If we think of taking the technology from a game like The Sims and deploying it for a serious purpose, such as a training aid for nursing, we have the potential to model and simulate, in game form, hospital operations and the like, providing an immersive experience for the nurse trainee.

Computer-generated story is the modeling of a story computationally such that we can build engines and tool suites that dramatically simplify the deployment of a new story for our networked game.

Modeling and simulation of human emotion is the frontier for networked games and simulations. For the entertainment world, the future of gaming includes developing an immersive gaming experience that has an emotional impact on the player. For the military, homeland security and defense, and hospital trauma worlds, we need a similar game-based simulation capability. The fundamental question is, How do we model human emotion such that we can author emotional experiences in game form in a controlled and appropriate manner? There are demanding requirements for such a capability across the spectrum of entertainment and serious-game developers, and it is critical that we perform the research needed to understand the potential human impact.

Understanding and analysis are a key element of any agenda for research into games. When humans are placed into large-scale MMOGs or into single-player modules, the question becomes, What happened during game play? What was the impact on the player? Current serious-game usage and large-scale simulation require human monitors to watch networked play. At the end of play, the human monitor comes back and says "I believe this team won, and here is why." We need an automated understanding and analysis capability for MMOG play such that we get a high-level report on what happened during game play over a specified period of time, from a particular viewpoint, with the ability to query that system for additional detailed information on why it reported as it did. There are defense, homeland security, and educational applications that require such automated analyses if we are to extend gaming much further into the serious-game domain.

Pedagogy/story integration is the insertion of pedagogy into a story, such that the story is immersive and entertaining, with pedagogy remaining subordinate to story. The game industry has experienced the failure of Edutainment, where educational software was sprinkled lightly with gamelike interfaces and cuteness. The story must come first and we must then learn how to insert pedagogy into the story creation and development process in the interactive medium of games.

Immersion

The sense of presence in a game is called immersion. It includes the following:

- Computer graphics, sound, and haptics,
- Effective computing—sensing the human state and emotion,
- Presence, and
- Advanced user interfaces.

R&D is needed on the technologies for engaging the mind of the game player by means of stimulation, for developing theories of presence, and for computing to sense human physical state and emotion.

Research on sensory channels is fundamental to the science and technology of games. As we move toward more capable graphics engines, we need to know how to appropriately utilize that new capability for our serious games, and we need to generate new technology that can be put into the next-generation graphics chipsets that industry provides. Spatial and immersive sound are key components of whatever training and educational systems we build with gaming. Future engineering requirements and human performance engineering need to be advanced to make sure sound is deployed appropriately and usefully for our serious purpose. Cross-modal sensory conflicts are an area for research. Haptics is also key to the future of games. If we believe that the R&D work we are performing now will be used for the technology that the game industry deploys in 10 to 25 years, there is still much to be done to improve sensory stimulation.

Affective computing entails measuring the physical and emotional state of human beings and transferring it to computer software. In the next 2 years, low-cost sensors will be available that measure the emotional state of the human and provide that as input to the running game. Devices will be needed to read the sensors and input the person's emotional state to the game. The game will need to be able to use that state as one of many inputs and respond appropriately. We do not really know what this response will look like. We do not have good models of human emotion nor do we have good models of how our computer characters should react to such inputs. We do know that such inputs will have a major impact on both the entertainment games and the serious games of the future. We need to understand that impact and to engineer and author it in a careful and controlled fashion. This type of research effort has the potential to broaden the scope and genres of entertainment and serious games. We may get to the point where a videogame not only makes us cry but knows that we are crying.

Presence is immersive experience offered to the game player or virtual reality explorer. Whether we are building a virtual reality or a game, we are attempting to give the player the illusion that he is in a virtual world. We need to be able to engineer presence such that we can create the effect we want rather than just hoping that it will turn out as we wish.

Advanced user interfaces become key as we move from the standard desktop PC to the mobile platform. There is much to be gained by studying how the game industry has developed interfaces that are almost universal—for example, if you can play Quake, you can play Unreal Tournament. We need to understand interfaces from the game perspective if we are to make good progress in the deployment of serious games.

Serious Games

Serious games research and simulations for nonentertainment domains include

- Serious game development across all application domains—health, public policy, strategic communications, defense, training, and education;
- Human performance engineering; and
- Game evaluation.

Serious game development is a fairly new phenomenon in the game world, and if the proper research is conducted it has the potential to eclipse the entertainment world in size over time. Here we are building games that use entertainment principles, creativity, and technology to carry out a government or corporate objective. As we engage in serious game development we need to establish principles, processes, and procedures for such deployment—usually called human performance engineering. If training and education are the objectives of our serious game, then we need to understand how to use the creativity of the entertainment world and combine it with appropriate human performance engineering principles. It is for that reason alone that the first serious games should (in fact must) be constructed in carefully controlled university or laboratory environments.

REFERENCES

Zyda, M., and J. Sheehan, eds. 1997. *Modeling and Simulation: Linking Entertainment and Defense*. Washington, D.C.: National Academy Press.

Zyda, M., A. Mayberry, J. McCree, and M. Davis. 2005. "From Viz-Sim to VR to games: How we built a hit game-based simulation," *Organizational Simulation: From Modeling and Simulation to Games & Entertainment*. W.B. Rouse and K.R. Boff, eds. New York, N.Y.: Wiley.

B

Social Behavioral Modeling

Kathleen Carley

A behavioral model is a model of human activity in which individual or group behaviors are derived from the psychological or social aspects of humans. Behavioral models include a diversity of approaches. The computational approaches important from the DoD perspective are social network models and multiagent systems.

An important caveat has to do with cognitive models. Cognitive models focus on the way in which cognition works, including aspects such as information gathering, processing, and utilization. Developers often build these models using general cognitive frameworks such as Brahms, Soar, Act-R, or various neural network platforms. Such models have traditionally focused on the individual in isolation and have rarely been used to model social dependent aspects of behavior that occur when multiple individuals are together. The key exceptions here are the cognitive multiagent systems (described later). Cognitive modeling has been used to address a number of DoD concerns, ranging from in-depth models of specific foreign leaders to detailed models of a human for use in evaluating various weapon systems or communication tools.

Social network models focus on the way in which relations among actors, such as who knows whom, constrain and enable access to information and behavior and serve as a basis for power and prestige. Multiagent systems focus on the way in which social behavior emerges from the actions of heterogeneous agents. Additional features, applications, and the state of the art will be described for each of these approaches in turn.

SOCIAL NETWORK MODELS

Social network analysis (SNA) is a computer-supported form of statistical analysis, derived from graph theory, that focuses on relational data (connections among nodes) rather than attribute data (features of nodes). Social network models are simply models of social behavior that take such relations into account. They may be realized as computational multiagent models, mathematical models, regression models, or conceptual models. Rather than detailing these models, the focus here is on SNA and the way in which such analysis and such considerations influence models.

SNA has received a great deal of attention since 9/11. As a result, many companies and individuals are claiming that they have expertise in the area even when they have no formal training or background. Catch phrases for fighting terrorism—"Disconnect the dots" and "It takes a network to fight a network"—and for doing business—"It's not who you know but who or what who you know knows" and "Are you networking?"—have appealed to our imagination and raised awareness of SNA. Further, there have been successful applications of this approach. For example, social network information was used to locate Sadaam Hussein, and several of the tools have been used in various criminal investigations. Students use social network information in Friendster to vet their dates.

The Nature of SNA Models

"Social network analysis" is a common term of art used to capture the different types of analyses done in three areas: traditional social network analysis, link analysis, and dynamic network analysis. These areas vary based on the number of nodes (multimode) and links (multilink) in the network and the scientific traditions out of which they emerged. These and other differences are summarized in Table B.1.

Traditional social network analysis centers on relatively simple networks. A typical social network analysis works with a single network connecting people to people by some relationship (perhaps they work together). Analysts in this area primarily use computational techniques to statistically analyze these networks. This area has a long tradition predating World War II. It emerged from the social sciences, particularly from anthropology and sociology, and has now spread to organization science, economics, physics, and computer science. Much of the work in this area has focused

TABLE B.1 Feature Comparison

Feature	Social Network	Link Analysis	Dynamic Network Analysis
Entity studied	The network	A set of links	Either the network or a set of links
Multilink	One or two links	Many links	One or many links
Multimode	One or two modes	Many modes	One or many modes
Focus	Identify key actors and groups	Anomaly detection	Identify key actors and groups
Networks evolve?	No	No	Yes
Locates network elite?	Yes	No	Yes
Locates patterns of behavior?	No	Yes	No
Locates patterns across networks?	No	Needs work	Needs work
What evolves?	Nothing	Nothing	Agents, groups, and networks
Predicts and assesses individual behavior	Few behaviors	Many behaviors	Many behaviors
Predicts and assesses group behavior	Few behaviors	No	Few behaviors
Handles missing information?	No	Needs work	Needs work
Optimized search?	No	Yes	Sometimes
Locates groups?	Yes	Yes	Yes
Analysis of change	Qualitative	Assumes the future is the same as the past	Quantitative
Handles streaming data	No	Needs work	Needs work

on characterizing the size and shape (topology) of the underlying networks, identifying who stands out (which individuals because of their relations to others occupy key positions in the network), and how the structure of the network or an individual's position within it influences behavior. There are numerous SNA computational tools, ranging from network visualizers to packages for analyzing network data, and new ones appear daily.

Link analysis centers on discovering patterns by looking at the relations among entities. Analysts in this area use computational techniques to locate patterns and subgroups. This area has emerged largely from computer science, with particular attention to work in machine learning. Some of the roots in this area are in forensics. Extraction of links often requires massive data preprocessing or restructuring of databases (Goldberg and Wong, 1998). Advanced data-processing techniques are combined with machine learning to enable rapid database transformation and pattern extraction. Much of the work in this area has focused on the identification and recognition of patterns, data mining, and node identification. There are a growing number of tools, many of which are available on the Web. Common tools exist for doing a variety of tasks, including extracting links from databases (Goldberg and Senator, 1998) and texts (Lee, 1998) and analyzing the extracted links (Chen and Lynch, 1992; Hauck et al., 2002).

Dynamic network analysis (DNA) is an emergent field centered on the collection, analysis, understanding, and prediction of dynamic relations (such as who talks to whom) and the impact of such dynamics on the behavior of individuals or collectives (Carley, 2003). Analysts combine computational techniques, such as machine learning and artificial intelligence, with traditional graph and social network theory and with empirical research on human behavior, groups, organizations, and societies to develop and test tools and theories of relational enabled and constrained action. This area builds on social network analysis and link analysis and adds computer simulation to the mix to look at network evolution. There are a growing number of DNA tools, some of which embody most of the SNA techniques.

Application Areas

Essentially any problem in which there are relational data—data about whether entities of one type relate to other entities of the same or different type—can be addressed using network analysis. Social network analysis, link analysis, and dynamic network analysis all have their own strengths. Factors that determine the effectiveness of one or the other type of analysis include the scale of the underlying network, the completeness of the existing data, and the level or types of errors in the data.

Social network analysis is particularly useful for understanding the connections among political elites, identifying groups and cliques in organizations, understanding the flow of information, understanding disease propagation in a group, or identifying the elite or the isolates. Grouping algorithms are useful for breaking a large network into a set of subnetworks such that the members are either tightly connected to each other (e.g., they form a clique) or are similar (e.g., connect to the same others even if not to each other). Applications are generally done at the individual level—person to person. However, the same tools can be, and have been, applied to organizations. In general, for most measures, the assumption is that the user has complete or almost complete data, that the connections are all of the same type (e.g., who lends money to whom), and that the nodes are all of the same type (e.g., they are all people).

Link analysis is particularly useful for identifying anomalous patterns. Typical applications are locating money laundering profiles or other sequences of activities associated with specific crimes and locating groups of people who have special relations to each other.

Dynamic network analysis has a variety of applications over and above those afforded by social network analysis. These derive from the fact that some networks can be co-analyzed, such as social and knowledge networks. As such, the tools can be used to assess organizational health and adaptability, to assess whether the movement of personnel between ships might reflect movement of information among owners, to locate emergent groups in terms of both who talks to whom and what are they talking about, to identify points of influence, and so on. Moreover, the simulation component facilitates assessing network evolution and evaluating courses of action designed to alter networks. Illustrative applications include disease spread, change in beliefs, assessment of various isolating or information-spreading courses of action, team design and assessment, identification of points of influence, and location of emergent subgroups.

State of the Art

There are a large number of computational tools within network analysis. These have been developed in the United States and Europe, at universities and by private companies. Few of these tools are interoperable. Some of the basic measures are available as open source, and algorithms for most measures have been published, although not collected into a single compendium.

In general, network analysis tools are increasingly widely used. Most metrics have received some level of validation or verification. Many of the underlying algorithms have been optimized (although not all tools contain the optimized algorithms). In addition, there is a rapidly growing body of basic and applied research. From a defense standpoint it is important to note that there is a rapidly growing body of applications on both red and blue force assessment. There is a wealth of information on the interpretation limits of the various measures. However, this information has not been systematically collected and organized.

There are technical references; however, they are out of date (e.g., Wasserman and Faust, 1994). There is no excellent undergraduate textbook. Knowledge needs to be acquired through classes at universities, special short courses offered by key practitioners to industry, didactic seminars at major conferences, and articles in key journals. In addition, practitioners and researchers in each of the three areas tend to attend different core conferences and read different journals.

In general, in traditional social network analysis, the most advanced tools are for binary and summarized data. Across this area the visualization tools are in their infancy, with the most advanced such tools being for networks of less than 200 nodes. Many of the link analysis tools use hidden Markov models or Bayesian updating, both of which are well-understood techniques. The most advanced DNA tools build directly on SNA and/or link analysis. Across the board, most tools stand alone and are not Web-enabled. Only a few tools are relatively easy to integrate into other systems. Only a few of the existing tools have been tested and optimized for large-scale networks (at least 10^6 nodes). Issues of measure robustness and sensitivity to missing or erroneous data are currently being addressed by several active research programs.

Key Limitations

There are a number of technological and practical limitations. In many cases, however, there is ongoing research to overcome them. One difficulty in this area is that the basic measures are so easy and the promise so high that many people and many companies are claiming expertise despite having no training in the area. Basic metrics are now in use in a wide variety of applications; however, claims about applicability and interpretation of results are often inappropriate, there is a great deal of reinventing the wheel, and there are spurious claims of novelty for well-understood approaches.

One of the key limitations in all of these areas is visualization, which is in its infancy. There are many tools for representing graphs, each with its own unique features. In general, most of the visualization techniques do not scale well for networks with more than 200-300 nodes. Visualization is often used for interpreting the results of the network analysis. The problem here is that exactly the same network laid out in two different ways is likely to be interpreted in two different ways by the person examining the picture. This is particularly true when the interpretation is done by a novice.

Interpretation of network metrics is also difficult. Even when metrics are normalized, there is little readily available information to tell how important a particular score is. What most analysts do is use the numbers in a relative fashion, comparing within the same data set whether a node is higher or lower than another node on some metric or comparing the metrics related to two data sets. However, there are no absolute guidelines. Nor are there compendiums of typical values and ranges for common social networks.

Most of the standard graph-theoretic metrics scale as N and a few as N^2, where N is the number of nodes. There is a normalized form for most metrics. Thus, most metrics can be used easily on graphs of varying sizes. There are a few metrics, however, that have been touted as critical, such as the "betweeness centrality" measure, that do not scale well. To achieve greater speeds for these metrics a special-purpose graph metric chip might be needed. Most clustering algorithms and pattern-location algorithms also scale at best as N^2 or N^3. These algorithms are still fairly new, and research is under way to improve the scalability of the algorithms.

In general, these tools are data-greedy. There are no standard techniques for estimating missing data or the size of the network or for dealing with erroneous data. Within link analysis, there are insufficient techniques for reducing the amount of data needed for robust learning in machine-learning pattern-location algorithms. Currently, however, there is ongoing research on the robustness and sensitivity of the underlying algorithms and metrics.

A common problem with all of the group location algorithms is that they locate a set of discrete groups; that is, nodes can only be in one group. This is true for techniques using some form of clustering or blocking in social network analysis and for pattern-location algorithms in link analysis. Robust, scalable fuzzy group techniques that generate socially meaningful groups are needed.

Currently the tools assume that the data have been precollected and, in many cases, preprocessed into specialized forms. Over time, analyses are typically done on historical temporal data. With the exception of some link analysis tools, very few of the computational tools can handle streaming data in an automated fashion. Ideally, these tools would be linked to live data streams and so provide updated information on networks as they change. However, for that to be feasible extensive research is needed on (1) determination of the meaningful temporal chunks for presenting relational data, (2) algorithms for updating metrics based on new data, (3) automated tools for parsing streaming data, and (4) visualization of the dynamics.

Relational data can be collected in a variety of ways, from automated data capture of various relational data streams (e.g., transaction records), to direct observation, to questionnaires. There are two related issues. First, little work has been done on (1) network-based privacy, (2) node anonimization, and (3) deanonimization in networks. This is likely to be a growing issue as more companies, such as search engines and e-mail vendors, increasingly provide social-network-based services. Second, little work has been done on estimating critical gaps in the data collected, which would allow data collectors to know where to focus.

Finally, from a defense perspective, network analysis tools have to take into account multimode, multilink data. The social network in isolation is of little value for evaluating courses of action. More predictive power and more analyses are made possible as the multiplicity of modes and links increases. For example, estimates of an actor's power require an understanding of how the individual is linked to others, to issues, to resources, and so on. This being said, key areas that need work are linking social network data (actor-to-actor) to tasks or events and locations. Neither the geotemporal aspects of networks and nor their resource-task aspects are well understood.

MULTIAGENT MODELS

As previously noted, multiagent systems focus on the way in which social behavior emerges from the actions of heterogeneous agents. Multiagent systems (MASs) are computer-based simulation programs in which there are a set of actors (called agents), each of whom can take action. Overall results derived from such systems depend on the sequence of actions taken by the agents. The agents typically act in parallel but need not. The agents are typically heterogeneous, but need not be. The agents typically can learn but need not.

MASs are often described as bottom-up systems because the behavior of higher-order entities, e.g., groups or populations, are driven by actions at the agent level. This is in contrast to system dynamic models, which are often described as top-down and in which the behavior of lower-order entities, e.g., agents, are inferred from change at the top level. Both types of models fall under the rubric of complex adaptive systems—especially when learning or evolution is involved and three or more rules or equations result in nonlinear interactions among components.

MASs go by a variety of names, often indicating the type of system that they are. Common names are multiagent-based systems, complex adaptive systems, agent-based systems, multiagent network systems, multiagent dynamic network systems, and cellular automata. In some cases, the name used is broad and applies to tools that are not agent-based as well. For example, complex adaptive systems encompass a wide variety of techniques including, but not limited to, multiagent systems and system dynamic systems.

TABLE B.2 Common Differences Across Types of Multiagent Systems

Type of System	Number of Agents	Algorithm Type	Cognitive Sophistication	Social Sophistication	Grid Based
Multiagent cognitive	Few	Rules	High	Low	No
Multiagent dynamic network	Many	Equations + rules	Moderate	High	No
Cellular automata	Many	Equations or rules	Low	Low	Often
Multiagent rule-based system	Many	Rules	Low	Low	Yes

Types of Multiagent Systems

There are many types of MAS. It is possible to classify them in a number of ways—for example, by the number of agents, the basic type of algorithm, the cognitive sophistication of the agents, the social sophistication of the agents, and whether or not they are grid based. A few classes of systems illustrate these differences:

- Multiagent cognitive models (such as a multiagent Brahms model).
- Multiagent dynamic-network models (such as Construct).
- Cellular automata and multiagent rule-based systems (such as those in Swarm, Repast, or Mason).

Table B.2 summarizes these differences. One caveat is that within any class, the actual level of general realism depends on the degree to which the model is utilizing actual data and the detail inherent in the underlying algorithms. Another caveat is that in principle MASs are ubiquitously applicable to problems that involve two or more actors whose behavior depends, at least in part, on the behavior of the other or others. The exact area of applications depends on the type of MAS.

As noted, these classification factors are the common dimensions along which MASs vary:

- *Number of agents.* On the one hand, multiagent systems include models with between 2 and 10 very cognitively sophisticated agents performing very in-depth, knowledge-intensive tasks. In such models, interactions among agents are typically prescribed by protocols for interaction and hierarchical precedents for who does what. Such models are more common in computer science and engineering; illustrative models are those involving Brahms or Soar. On the other hand, a MAS may be made up of thousands or millions of cognitively simpler agents doing relatively simpler tasks. In this case, interactions among agents are the result of the agents meeting and greeting each other, trying to occupy the same space, and/or exchanging or consuming resources. Such models are more common in the social and organizational sciences, biology, and physics; illustrative models are those coming out of the Santa Fe Institute or the Brookings Institution.
- *Algorithm type.* In some MASs, the agents are systems of equations specifying the state of the agent and how it changes or learns as new information arrives. In such models, machine-learning and pattern-recognition software may be used to create adaptive agents. Some MASs of this type employ neural network technology or simulated annealing technology and so enable agents to act as heuristic-based optimizers. In other MASs the agents are a body of rules. In such models, expert-system and pattern-matching software may be used to enable dynamics. In these MASs the rules rarely change over the course of a simulation unless a heuristic optimizer controls the simulation and forces rule change through either automated subgoaling and rule construction procedures (as in Soar) or through the use of a heuristic-based optimization procedure such as a genetic algorithm.
- *Cognitive sophistication.* In general, the cognitive sophistication of the agents is inversely proportional to the number of agents. Thus, you are likely to see an MAS with a few very sophisticated agents and an MAS with many cognitively trivial agents. When the agents are cognitively sophisticated, the cognitive model often includes features such as recognition, planning, memory, and decision-making modules. Models are often built using a handful of actors, each built in one of the common cognitive modeling platforms such as Brahms, Soar, ACT-R, or neural nets. In such cases, the agents may have features that enable them to forget, make mistakes, and create new modes of behavior. The most cognitively sophisticated MAS are those that use an underlying cognitive modeling architecture such as Brahms or Soar. Models written directly in a high-level language such as C++ are likely to be moderately cognitively sophisticated, whereas those written in an MAS framework such as Swarm, Repast, or

Mason are typically cognitively extremely simplistic. Models that opt for large numbers of cognitively simplistic agents argue that social processes and complexity are an emergent property of interaction among large numbers of simple heterogeneous agents.

- *Social sophistication.* Most multiagent models are extremely unsophisticated socially. That is, rarely do such models take into account the impact of sociodemographic characteristics, social networks, or interaction with social groups and organizations. Typically, in MASs with a few cognitively sophisticated agents, social factors are either ignored or prescribed in terms of a communication and command hierarchy; as such, social behavior is constant. In contrast, most MASs with millions of cognitively simple agents, particularly those built in the MAS frameworks, do not model real social networks or groups but may differentiate actors on anywhere from two to five sociodemographic dimensions. There is a new class of models, however, the multiagent dynamic-network mode, in which networks such as social, knowledge, and resource networks enable and constrain interaction among agents and the networks coevolve as the agents interact.
- *Grid based.* Most MASs with vast numbers of agents have the agents operate on a grid. There are two forms of grid-based models. In the first, each point in the grid is an agent, and the agent's health and action are a function of the health and actions of nearby agents. In the second, cells in the grids are locations through which agents move (right-left, up-down), where they consume or leave resources and interact with the agents they meet in the same or neighboring cells. The classic example of a grid-based MAS is the game Life. Today, many MASs based on grids are barely more complex than the original Life system, although modern systems use a toroid rather than a strict grid to avoid edge effects. The MASs with a few cognitively sophisticated agents typically do not operate on a grid. Rather, if they need location they treat location as a variable in the rules that they use to operate. Grid-based MASs are in sharp contrast to dynamic-network MASs, in which the agents operate in an ever-changing social space where "nearness" is defined on the basis of social, cultural, political, knowledge, or task factors. In this case, if location is needed, it is often treated as proximity and is just one of many factors defining the nearness of two agents and their propensity to interact.

MAS Toolkits

There are a number of MAS toolkits currently available. These toolkits are a framework language in which to build an MAS. They facilitate system building because they already have built-in procedures for common functions such as displaying agent interaction, displaying change in variables over time, garbage collection, general input/output, and some statistical procedures. In some cases, the toolkits are made available along with sample agents. In general, from a learning perspective, such toolkits work well in the classroom because they reduce time spent on extraneous factors and let the novice quickly build a prototype system. However, from a deployment perspective, these frameworks have some drawbacks for anything but proof-of-concept systems. In general, MASs built in these tools are slow, and often better optimization can be achieved by writing in languages such as C++. The frameworks are best suited for many agents at the same level of granularity. Thus if you want agents representing humans to interact with agents representing companies or with agents representing institutions, the basic communication, learning, and behavioral features are not available. Such multigranular models can be built in these frameworks; however, it is often more complex than building them directly in an object-oriented language. Most of the toolkits do not have drill-down explanation facilities. Increasingly, these toolkits are making it possible to run the simulation in a Monte Carlo fashion and extract statistical properties; however, the toolkits rarely export data in a form readable by standard statistical packages, data-farming environments, or response-surface analysis tools.

Some of the MAS toolkits have regular user groups, training seminars or courses, and online help. Some of the tool kits are open source, such as Repast; others are held by companies (e.g., Swarm) or universities (e.g., Mason). In general, translators from one toolkit to another or from one language (such as C++) to a toolkit do not exist. Consequently, it typically takes about 75 percent of the time it took to develop the original system to rebuild it in the toolkit, assuming that the original system had moderate documentation. The concept of toolkits for MAS is a powerful one. Today, however, the extant toolkits are still in their infancy.

State-of-the-Art Applications and Limitations

The value of any simulation, including MAS, is partly tied to the level of realism in the model. Any simulation system is a model and so should be less complex than the real world; however, oversimplification results in models that are so high level or so incorrect that the results can be overinterpreted or misinterpreted and so should not be used for policy setting and decision making. The rule of thumb is to make the model only as complicated as it needs to be to address the issue of concern and the necessary level of fidelity.

In MASs, adding more rules or equations increases the realism of the system and its usefulness for decision making. Opponents often argue that the more equations or rules, the worse the model. Arguments include appeals to parsimony, Occam's razor, or understandability. A typical argument is that as the model increases in complexity (number of variables and rules/equations) it becomes increasingly likely that the model can be made to fit any possible outcome. This argument derives from econometrics, where as the number of variables approaches the number of cases, the underlying data can be completely and perfectly modeled. This argument, however, is not directly applicable to MAS. In MAS, the addition of new rules and equations serves to increase the number of outcome, or dependent, variables that can be generated rather than, as in econometrics, the number of independent variables, whose relation to dependent variables can be explained. Further, in an MAS, the rules and equations are effectively a multiple constraint set, which reduces rather than increases the number of outcomes a MAS can generate. A side product is that the addition of empirically based rules and equations often increases the plausibility of the results generated by the model by reducing the space of implausible results.

The validation of MASs is a complex issue worthy of several volumes. Rather than trying to review all aspects of validation, only a few high-level points will be made. First, most MASs are never, and probably never should be, validated. The simpler the model, the less likely that it can be meaningfully validated using techniques other than generic face validation. The level of validation required of such models depends on their purpose. If the purpose is to demonstrate a proof of concept, or that something is possible, then minimal, if any, validation is needed. Face validation typically suffices. Second, MASs are difficult to validate in full and are generally validated only within a small area of performance. A typical approach to validation is to run a virtual experiment using the MAS, take the generated data, statistically analyze the results to generate the response surface, and then contrast the response surface with real data. Since it is easy to generate so much data that no existing statistics package can handle them or so much data that most desktops cannot store them, only small portions of the overall response surface can be estimated at once. The size of the analyzed response surface is often dictated by the user's interests, critical policy or decision-making questions, the storage capacity of the machine doing the analysis, the data capacity of the statistical tool, and the time it takes the simulation to produce the necessary data. Third, MASs are difficult to tune and validate as changes in one part of the system have unforeseen effects on other parts. As noted, a MAS can be thought of as a set of mutually constraining and interacting forces. As such, a change in one component often necessitates the revalidation of earlier validated or tuned components. For some systems, intelligent software systems are needed to do the validation of the MAS. Finally, MASs are often difficult to validate, as the necessary real-world data may not be available.

The realism of MASs can be increased and their value to DoD increased when they are linked to real data. Most groups that build MAS systems have contrasted, at best, the results of one dependent variable with real data. Only a few systems, such as some recently created for the Defense Advanced Research Projects Agency or the BioWar system, use massive amounts of real data to set the input specifications of the models and other data to validate the system. In general, this means linking MASs to database systems. The key technical challenge here is that as the ontology in the database changes, the MAS needs to be augmented. There are currently no tools to facilitate such changes. A second challenge is that, for validation, it is important to have the MAS produce data in the same form as the real data—that is, create a comparable database. There are currently no standardized tools for doing statistical comparison of data in two identically structured databases.

MASs using cognitively sophisticated agents tend to require the use of knowledge engineering techniques. Such models tend to be special purpose and have minimal reuse. Their key value is to take the place of human teams in wargaming situations and in equipment testing and design situations and to evaluate processes that facilitate team behavior. In general, these models use various cognitive architectures with multiagent components added and so are often limited to only a small number of agents. Their strength is looking at detailed task-related behavior. As previously noted, such models tend to use predefined social interactions. This limits their use in war games as the MASs do not adapt the interaction process but just the task-based communications and actions.

Typical grid-based MASs with millions of cognitively unsophisticated agents are generally useful only for high-level explorations of general concepts. They are valuable for starting groups to think outside the box and for provoking discussions. These models are rarely sophisticated enough to be used as an adaptive adversary in war gaming or for evaluating task-based behavior. The strength of these models is their ability to look at population-level trends resulting from local action. As such, they show promise in areas such as marketing, determining the impact of psychological operations, information diffusion studies, and disease transmission studies. Rarely do such models generate actionable intelligence.

Now consider multiagent dynamic-network systems when they are tied to empirical data. Such models utilize agents with moderate levels of cognitive sophistication and high levels of social sophistication. This makes them useful for war gaming to look at adaptive adversaries. Given current technology, this combination results in models that can handle more agents than the cognitively sophisticated models but that run more slowly than the grid-based, cognitively simplistic models. As such, the strength of these models lies in representing and reasoning about reasonably large populations. The added cognition and social sophistication inherent in these models makes it possible to produce actionable results. However, getting a model to the point of producing actionable results takes a multiperson, multiyear data collection effort on top of a multiyear model development effort.

One of the key factors limiting MAS models from a DoD perspective is the modeling of action. Currently, actions can be modeled at a very high level (pro-con, hostile, friendly, neutral) or at a very detailed level (fire a particular weapon). There is neither a middle ground nor a hierarchy relating actions at one level to another. MASs that try to model actions tend to be either very generic or single-use. A basic ontology of actions is needed for the state of the art to advance.

Development of an MAS

It is important to note that, with MAS toolkits, even bad programmers and novice simulators can build interesting and seemingly powerful MASs. Such models can be built in the course of a few months. As a result, we are now seeing thousands of small systems being built by individuals or small teams. For example, individual soldiers with little or no training in simulation are now building MASs and using them to inform critical decision making and policy. Thus, the use of MASs enables the analyst to systematically consider the interaction among more factors and so to base decisions on a more thorough analysis. However, the development of MASs by those not trained in simulation means that the results of the systems are often misinterpreted, and classic mistakes are often made that cause the results from the model to reflect incorrect simulation practices rather than interactions among the factors modeled.

Very detailed sophisticated models that produce actionable results often need to be developed by a team working collectively for 3 to 5 years. It makes sense to use separate teams for data gathering, validation, and usability testing as each of these areas requires different types of scientific skills. In addition, the team building the model often needs to employ many of the same techniques for development that are used in system engineering.

THE WAY AHEAD

Key advances and applicability to defense modeling require that MASs and network analysis techniques be integrated into tool chains. For example, pattern-discovery techniques can be used to derive equations from historical data that can then be used in MASs to evolve future systems. MAS techniques can be used to evaluate causes of action and suggest areas for further SNA data collection. Combining these techniques will enable new types of problems to be solved; for instance, combining social network metrics with a pattern-discovery technique is key to building an understanding of how networks grow and evolve.

This is not to suggest that DoD should move to a large integrated behavioral model—quite the contrary. The development of MAS frameworks and the explosion of network analytic tools is making social behavioral modeling widely available and is leading to the development of many small, single-purpose tools. If they are to be fully explorted, they need to be made interoperable. It is important to note that it would not be feasible to require all tools to be written in a single language or to use a single framework; rather, the solution will be to integrate models not only from diverse domains but also in diverse languages. Multiple models and visualization tools should be available to address diverse problems, but in a way that data (real and virtual) can be easily shared among the various tools. A variety of things are needed to support such interoperability, including standards for the interchange of relational data. Behavioral modeling tools need to be Web-enabled, and XML IO languages must be developed, along with a uniform vocabulary for describing relational data, which is particularly critical as the tools and metrics are coming out of at least 20 different scientific fields.[1] For defense and intelligence applications, we need to further explore common platforms and data-sharing standards so that tools written in the unclassified realm can be rapidly moved, without complete redesign, to the classified realm. Interoperability and common platforms and ontologies for these tools will enable novel problems to be addressed more rapidly by regrouping existing models and they will enable various subject-matter experts to interact through their models, thereby allowing a broader approach to problems, reducing the likelihood of a biased solution, and facilitating rapid development and deployment.

Current tools are either very data greedy or become more valuable as they are linked to real data. However, there is a dearth of relevant data currently available in clean, preprocessed form. To reduce the time spent by analysts in data collection and increase the time spent in analysis, automated and semiautomated tools for data gathering, cleaning, and sharing are needed. Such tools should include natural language processing tools for extracting relational data from audio and text sources, Web-scraping tools, automatic ontology generators, and visual interpretation tools to extract network data from photographs and visual images. Appropriate subtools for node identification, entity extraction, and thesaurus creation are also needed. The development and availability of these tools in an interoperable environment are critical for providing masses of data that can be used for model tuning and validation. More rapid data collection would also mean the availability of more data sets for doing meta-analyses, thereby enabling the theoretical foundations of the field and our understanding of social behavior to be improved. Finally, these tools are essential to providing the wealth of data needed by the social behavioral modeling tools in order that the models make reasonable forecasts or provide reasonably accurate analyses of situations and organizations.

---

[1]These fields include anthropology, sociology, psychology, organization science, marketing, physics, electrical engineering, ecology, biology, bioinformatics, health services, forensics, artificial intelligence, robotics, computer science, mathematics, statistics, information systems, medicine, civil engineering, and communications.

Improved speed for many of the algorithms could be provided by computer architectures designed for relational data or by the use of special integrated circuits with embedded versions of the less scalable algorithms. This would enable a speed savings beyond that afforded by current vector technology. Such technology would facilitate faster processing and enable more real-time solutions, particularly for large-scale networks.

To reduce the qualitative aspect of interpretation in this field, a living archive of collected network data is needed, replete with information on metrics for the nodes in each data set. Such an archive could be used to set context information. For example, such information could be used to evaluate whether the density of particular networks is exceptionally high or exceptionally low, or whether values for connectedness of individuals are out of range. Such an archive would facilitate meta-analysis and comparative analysis. This is critical to increasing the theoretical foundations of the field and improving our understanding of social behavior.

MASs designed for applied settings need to be placed in data-farming environments. These environments need to be augmented with special-purpose tools for running massive virtual experiments, improved visualization and analysis tools, and semiautomated response surface generators. Current data-farming tools are often cumbersome to use, require code modification of the MAS, or are limited by the processor speed and storage capabilities of the machines that they run on. In order for MASs to be routinely run in data-farming environments, new, more flexible environments need to be developed and made easily available to analysts using MAS; and MASs need to be developed with wrappers so that they can be placed in these environments. Standardized input/output formats need to be developed. By routinely placing a MAS in a data-farming environment, a better understanding of the range of possibilities forecast by the model will be derived. This will enable the MAS to better support policy and decision making.

Currently, when MASs are used to inform policy and critical decisions, the models are often run only a few times in carefully controlled virtual experiments. While this approach enables the analyst to explore more possibilities more systematically than not using a simulation at all, errors could still be made if the results are interpreted beyond the scope of the experiment. By placing the models in a data-farming environment, the number of virtual experiments considered, the range of possibilities examined, and the scope conditions analyzed can be expanded often by several orders of magnitude, providing a stronger basis for decision making. Further, once a model has been validated, the response surface equivalent can be used as a rapid model in training situations where the users don't have time to wait for the MAS to finish running.

Another avenue that is likely to promote major breakthroughs is the linkage of social behavioral modeling to gaming environments, particularly online multiplayer games such as Everquest and America's Army. Major research initiatives are needed to explore the link between social behavioral modeling and gaming tools. Possible research areas are the realism of the social behavior exhibited in these models, the use of the MAS to provide flexible opponents and/or to make the apparent number of game players larger and so force players to think about group scale issues, the ability to track and analyze behavior using dynamic network analysis techniques, and the use of these games to generate data to test tools. Key benefits here would be improved training tools and visual what-if scenario evaluation.

As previously noted, there needs to be additional development in a number of areas, including attachment of models to streaming data, improved visualization, metric robustness studies, and so on. Progress here will require linking social networks to other types of data such as location and event information and linking diffusion theory to other forms of theory such as action and cultural theory. This will require funding both basic and applied research. It will also require increased recognition for and acceptance of applied social science research in universities. Currently there are a number of funded research efforts in cultural modeling, geospatial link analysis, and adversarial modeling, all of which are supporting work along these lines, much of it directed at providing usable systems in several years. This is a positive development, particularly when such modeling efforts are based on strong empirical and theoretical foundations. However, although much basic research remains to be done in developing a task ontology, a unified model of culture, or even a shared definition of culture, relatively little research funding is being directed to it. The key here is not to simply invest in the social sciences but to invest in the mathematical and computational social sciences that will ultimately support defense needs. One benefit will be an improved understanding of basic social and cultural phenomena. Another benefit will be a decrease in misleading models that appear to be social but are not theoretically or empirically sound.

At the same time, most of the research community, particularly in the social sciences, is not focusing on applications. The mere idea of hard deliverables, common practice in engineering and computer science, is contrary to the culture of most social science departments. Thus while there is a strong need for quantitative social science modeling on defense issues there is a dearth of highly trained social scientists involved in applied work. Universities need to expand their undergraduate social science curriculums to include more of the mathematical and computational social sciences. In particular, undergraduate courses should routinely include social network analysis, basic simulation, and multiagent systems. Universities need to encourage and facilitate applied research and to adopt engineering-style curricula focused on social and policy applications. Master's programs that combine social and computational science need to be developed. Military universities such as West Point and the Naval Postgraduate School should also offer social network and MAS courses. The development of such curricula and degree programs is vital to our national intellectual strength if we are to remain at the forefront in this area and to have a stronger workforce of computational social analysts capable of developing and using social behavioral models.

Analysts engaged in social behavioral modeling who are trained in computer science, engineering, or physics should work in teams with social scientists to avoid reinventing the wheel or making common-sense assumptions about social processes that have no empirical basis. Corporations need to provide time and resources for selected personnel to become jointly trained in computer science and social science either by increasing the number of personnel sent to master's programs, bringing in relevant faculty to teach short courses, or engaging in more joint research with universities as equal partners contributing the missing skill, social or computational. The key advantage of teaming is that it will improve model development and will serve as a stopgap until more computational social analysts are trained.

Expected Outcomes

Success in the activities outlined above would facilitate the rapid development and deployment of social behavioral models that allow systematic reasoning about various courses of action in a wide range of realms. More courses of action could be evaluated in less time and more systematically than is done with conventional tabletop war gaming or current non-computer-assisted analysis of relational data. Such models would also reduce time spent in data processing and increase time spent in analysis and interpretation. They would facilitate what-if analysis and could ultimately support near-real-time, what-if analysis in the field. This would be a clear force multiplier.

These activities would increase the maturity of this field, improve scientific theory, facilitate rapid linking of models to solve novel problems, and encourage new discoveries. They would promote the development of a new science that combines computation and society, just as the previous combination of computer science, design, and psychology led to the new science of human-computer interaction.

REFERENCES

Carley, Kathleen M. 2003. "Dynamic network analysis." *Dynamic Social Network Modeling and Analysis: Workshop Summary and Papers*, Ronald Breiger, Kathleen Carley, and Philippa Pattison, eds. Washington, D.C.: The National Academies Press, pp. 133-145.

Chen, H., and K.J. Lynch. 1992. "Automatic construction of networks of concepts characterizing document databases." *IEEE Transactions on Systems, Man and Cybernetics* 22(5):885-902.

Goldberg, H.G., and R.W.H. Wong. 1998. "Restructuring transactional data for link analysis in the FinCEN AI system." *Proceedings of 1998 AAAI Fall Symposium on Artificial Intelligence and Link Analysis*. Menlo Park, Calif.: AAAI Press.

Goldberg, H.G., and T.E. Senator. 1998. "Restructuring databases for knowledge discovery by consolidation and link formation." *Proceedings of 1998 AAAI Fall Symposium on Artificial Intelligence and Link Analysis*. Menlo Park, Calif.: AAAI Press.

Hauck, R.V., H. Atabakhsh, P. Ongvasith, H. Gupta, and H. Chen. 2002. "COPLINK concept space: An application for criminal intelligence analysis." *IEEE Computer Digital Government Special Issue* 35(3):30-37.

Lee, R. 1998. "Automatic information extraction from documents: A tool for intelligence and law enforcement analysts." *Proceedings of 1998 AAAI Fall Symposium on Artificial Intelligence and Link Analysis*. Menlo Park, Calif.: AAAI Press.

Wasserman, S., and K. Faust. 1994. *Social Network Analysis*. New York, N.Y.: Cambridge University Press.

C

Composability

Paul Davis

INTRODUCTION

Composability is the capability to select and assemble components in various combinations to satisfy specific user requirements meaningfully. In modeling and simulation, the components in question are themselves models and simulations. Although terminology on such matters varies, the committee distinguishes composability from interoperability by the requirement that it be readily possible to assemble components differently for different purposes. Interoperability, on the other hand, may be achieved only for a particular configuration, perhaps in an awkward one-time lash-up. To put it differently, composability is associated with modular building blocks. It is also useful to think of composability as a property of the models rather than of the particular programs that happen to implement them in particular ways.[1]

A recent monograph reviewed the issues of composability in some depth (Davis and Anderson, 2003). It identified and discussed the impediments along four dimensions:

- Complexity of the system being modeled.
- Difficulty of the objective for the context in which the composite M&S will be used.
- Strength of the underlying science and technology, including standards.
- Human considerations, such as the quality of management, having a common community of interest, and the skill and knowledge of the workforce.

The monograph had recommendations for actions in each category, but that material will not be repeated here. Instead, this discussion builds on the earlier work.

[1]These definitions and distinctions are based on suggestions drawn from several recent studies—Petty and Weisel (2003); Davis and Anderson (2003); and Page et al. (2004). Although by no means universal, they appear to be sound and quite useful.

DISCUSSION OF SELECTED ISSUES

This section, which makes up the remainder of Appendix C, provides somewhat extended discussion of particular important topics meriting attention. It draws on the recent literature, but offers some new ideas as well.

Focusing on Challenging But Feasible Objectives, Not Grails

As discussed in Chapter 1, the committee recommends moving away from the plug-and-play ideal toward a more feasible but still challenging definition and conception of composability. It is sometimes argued, however, that establishing ideals and then moving toward achieving them is useful even if the ideals are unachievable. We believe that to be a dangerous strategy for the DoD as it seeks to promote composability. Among the problems with such an ideal-focused strategy are the following:

- Those writing proposals to the government tend to promise to deliver what the government demands, even if they know better. The proposal that "stretches" the most (i.e., is the most unscrupulous in its promises) may win because those reviewing the proposal see lesser proposals as insufficiently ambitious or as insufficiently responsive.
- Those developing the models will have strong disincentives for being clear about the shortcomings of their approach. Thus, they will emphasize their delivering composability when in fact what the are delivering will provide "engineered composability" within a narrow domain. The composability problem may seem to have been solved, when that is most certainly not the case.
- Because the shortcomings are submerged, higher-level managers may not even recognize the need to be investing in methods that would mitigate the problems.
- Overall, a kind of intellectual corruption can set in, in

which public comments generally (and even, sometimes, informal discussions) skirt around known problems, to the point at where the proverbial emperor's lack of clothes is not even consciously recognized.

In time, such problems are usually resolved because people seek to do a good job, and facts do matter. Moreover, it is natural for scientists and engineers to be questioning and assertive. It is gratifying to note that the technical literature now has numerous candid and thoughtful articles on the subject of composability, articles that go well beyond what they did a few years ago. Nonetheless, recovering from the promulgation of poorly conceived goals and initiatives can take many years.[2]

Syntax, Semantics, Pragmatics, Assumptions, and Validity

A continuing difficulty in the discussion of composability is distinguishing among the kinds of problems that arise. The usual discussion still refers to problems being either syntactic or semantic, but the situation is more complex than is conveyed by that description. The following items provide a tutorial and recommend distinctions that need to be made systematically. They deal with syntax, semantics, pragmatics, assumptions, and validity.

- *Syntax.* In shorthand, consistency of syntax means that two models can operate together. That is, the digital output from one can be read as the digital input to the other. Protocols such as HLA are designed to assure syntactical consistency among models to be connected.
- *Semantics.* Semantics is usually defined as "meaning." Thus, if some data can flow from Model A to Model B, the semantic question is seen as whether those data are understood by both A and B to mean the same thing (e.g., the current personnel strength of a battalion). To computer scientists, however, the operational meaning of semantic consistency is often much narrower and computer oriented.[3]
- *Pragmatics.* Consistency of meaning is not always straightforward because the same word means very different things depending on context. Moreover, key aspects of that context may not be explicit. This is the realm of pragmatics. The term "force ratio," for example, may refer to the ratio of forces as measured at a theater level, an operational level, or a tactical level. Even if one knows that the tactical level is the one intended, the term remains ambiguous because it can refer to battalion-level conflict or something more microscopic, such as when individual fighting vehicles and infantry are engaged. The related force ratios are not the same. Another aspect of pragmatics involves ontology. One model may have a built-in concept that a squad is a component of a platoon, which is a component of a company, and so on. Another model may have no such assumed structure because, in the context for which it was built, such an assumption was not necessarily correct (as in a treatment of special forces operations, which often involve very small teams that are not effectively associated with higher units during their operations). For the two models to compose well, it may be necessary for any such discrepancy to be resolved.[4]
- *Assumptions.* The difficulties continue. The data's meaning may be well understood, but the way Model A calculates the data may not be suitable for what B needs. Sometimes this may be a matter of precision, but other times it may be considerably more subtle. "The temperature," for example, might refer to a surface temperature, an average temperature over some path into the ocean relevant to a sensor, the ambient air temperature on a battlefield with very hot moving objects, etc. Still other times, the calculation reflects assumptions that are only sometimes valid. This is more than pragmatics as that term is usually used in linguistics; it involves "assumptions."
- *Validity.* And, finally, there is the question of whether the assumptions are correct. If composability includes the requirement that a composition be meaningful, that

[2]We mention three examples, recognizing that each initiative had its defenders. The examples are (1) DoD's heavy-handed and oversold mandating of ADA in the 1980s; (2) its introduction of the very useful high-level architecture (HLA), deferring work on the harder issues discussed in this report; and (3) its mandating of centralized development of "ultimate models" for training and analysis (JSIMS and JWARS), accompanied by the often-expressed view that other models would and should disappear. ADA proved useful only for niche work, because of widespread adoption by industry of other languages (e.g., C++). HLA has proven successful for promoting interoperability of legacy models but was never a solution for the challenges of semantics, pragmatics, and validity discussed here. The JSIMS program was cancelled after programmatic failures. The fate of JWARS is yet to be determined, but it seems likely that at best it will have a useful niche role in some aspects of analysis. It is unsuited to capabilities-based planning under deep uncertainty and the agile analysis of transformational concepts, points recognized nearly a decade ago (NRC, 1997).

[3]Examples of these relatively straightforward semantic issues are lexicographic problems—i.e., the syntax may work, but the usage cannot make sense. Some homely examples are division by zero, providing an alleged value of an array that is inconsistent with the array's dimensionality, or using a character not permitted by the language.

[4]The issue of ontological assumptions is emphasized in recent work by Andreas Tolk and students (Turnitsa, 2005). Other authors regard such matters as more a matter of assumptions than of pragmatics, which they see as contextual meaning (Davis and Anderson, 2003; Hofmann, 2004). Most software engineers and computer scientists subsume pragmatics, assumptions, and validity under pragmatics (see, e.g., Szyperski, 2002).

would seem to require adequate validity. Underlying assumptions, however, may be consistent but wrong.

The distinctions we have noted are closely related to, but not identical with, the conceptual levels of interoperability defined and discussed by Tolk and students at Old Dominion University (Turnitsa, 2005). Those range from no interoperabilty through physical, syntactic, semantic, pragmatic, dynamic, and conceptual, with the latter corresponding to complete substantive agreement between the models in question.

Separating Conceptual Model, Implemented Model, Simulator, and Experimental Frame

Background

The desirability of distinguishing between a conceptual model and a program representing a particular implementation of that conceptual model has been emphasized for decades by thoughtful scholars such as Ziegler et al. (2000) but ignored by the vast majority of model builders, who leap directly into programming and often leave behind very little that might pass as respectable documentation. The result is not only a product that is difficult to understand or modify, but one that is linked in subtle and sometimes insidious ways to the particular implementation environment (programming language and simulator). Furthermore, the result is often not well designed because it was not adequately reviewed and iterated at an abstract level, where design plays such an important role. These matters are being increasingly appreciated, as reflected in the success of the Model Driven Architecture (MDA) effort and some modern textbooks that teach design and clarity of thought while remaining practical (Blaha and Rumbaugh, 2005).

Rethinking the Issue in the Light of Conflicting Considerations

There are, then, strong reasons for advocating separation of conceptual model, implemented model, simulator, and experimental context in model development and usage. At the same time, there are strong technological pressures working in the opposite direction. No one today would think of working out the detailed specifications of a model on a typewriter, to be handed over subsequently to a programmer to implement. Even those who favor designing in UML sometimes have mixed emotions because, in practice, so much is learned by iteration—early design notions represented in UML may prove foolish when someone gets into details. If the same person is designing and implementing, the iteration may be easy, but if formalized separations exist, then the person closest to the code may have to go through what amounts to an appeals process in order to change the design reflected in the UML. That may be good in the sense of maintaining discipline and avoiding ad hoc changes, but it may be bad in the sense of delaying and obstructing important improvements.

The tensions between top-down and bottom-up are longstanding and will surely continue. They are accompanied today by the reality of horizontal, distributed collaborations and by the improved efficiency of integrated environments that provide tools for everything from diagram sketching through programming and statistical analysis of simulation outcomes. It would seem most unwise (and probably most unfruitful) to argue that the strict separations suggested by the older academic writings be reimposed. What, then, might be done?

Tentative Suggestions

If we rethink what the purposes of the separations are, the desirable path becomes clearer. Those purposes include the following:

- A conceptual model should exist because it displays the big picture, the design, and the linkage between application and model. It is the conceptual model that can best be communicated, in different forms, to clients, to other modelers, or to those concerned with composition. Realistically, the existence of good documentation depends on existence of a conceptual model.
- The virtues of a conceptual model disappear if it is cluttered by implementation details.
- The ability to comprehend a conceptual model and conceive of alternative implementations depends upon the conceptual model being expressed in implementation-independent terms.

One way to deal with such considerations is to develop and maintain a rigorously independent conceptual model, such as might be expressed in UML diagrams augmented with other methods. However, no one who has worked with higher-level modeling environments such as Mathematica, Analytica, iThink, or Extend would regard that approach as the only way, or even necessarily the best way. Suppose, for example, one had designed a model using one of these systems. One would need all or most of the following: a visual representation of the model, a hierarchical text-based representation, a clear list of inputs and outputs, definitions, and probably many notes. If one wanted to implement the model in some other system, it would often be rather easy to do so. This would require recoding, not merely sending the electronic files from one system to another; coding itself, however, is not so time-consuming as are conceiving and designing.

What matters most, then, is that the conceptual model can be viewed and comprehended separately, without being caught up in the details of implementation. If that conceptual model happens to have been developed simultaneously with implementation, that fact does not materially interfere with the purpose of the conceptual model. Moreover, having proven the feasibility of the conceptual model with an implementation has great advantages.

As for having implemented the model in a particular language, that may or may not be a problem. Higher-level languages such as that in EXCEL, Analytica, etc. are for the most part usable as a kind of pseudocode when discussing the conceptual model. The clutter associated with the programming detail can be suppressed when doing so.[5]

Simulators. Separating the simulation model from the simulator is important because simulators often have built-in limitations that affect validity of the results. This could be something obvious, such as the time step permitted in time-stepped simulation, the inability to vary the size of time steps dynamically, or the inability to implement discrete-event simulation. It could be more insidious, however, such as when the "simulator" deals not only with time, but also with terrain and environment, in which case the "simulator" is actually modeling part of the system, perhaps in ways that constrain or override intentions of the conceptual model. This can not only frustrate intentions of the designer but also make verification and validation extremely difficult.

It seems, however, that the built-in simulator function of an integrated modeling environment should not be seen as particularly troublesome, because if its approach to simulation (e.g., continuous rather than discrete-event) is a problem, that will likely be evident and the developer can choose to reimplement the model once its basic design is frozen. Other reasons for doing so might also exist; they might involve efficiency, interoperability with other models, and so on. In contrast, it is indeed troublesome if a conceptual model has been implemented in a system that somehow locks in a particular concept of terrain (e.g., grids versus hexagons versus a vector approach), command and control, or other substantive features of the real world. Thus, special care should be taken to separate the conceptual model from those implementation-specific features. How to accomplish that in general is not clear, but the entanglement of models with their programming environment's infrstructure (e.g., its treatment of terrain) can be very troublesome to composition efforts (Hofmann, 2004).

Experimental Context. The concept of explicitly defining the assumptions, purpose, and plan of analysis in an experimental plan (what is often referred to as defining the experimental frame) is very important.[6] So also it is important to distinguish this aspect of the MS&A effort from model development, although thinking through use cases and the anticipated-and-plausible analytical requirements can strongly affect development. Since the tools available in the MS&A community are not generally well developed on this matter workers tend to do much of what is required "on the side," perhaps using EXCEL or some special scripts to help themselves organize and drive simulation and subsequent analysis. An exception to this, which can be seen as an existence proof, is Mathematica. Those who use Mathematica are able to write text, design, program, simulate, analyze, chart, and record without leaving the Mathematica environment. Further, they can choose to some extent how they wish to conduct dynamic simulation and they can call upon a wide range of library functions, many of them subject-area-specific (economics, physics, and so on). There are both advantages and disadvantages to such an approach, but the advantages are considered persuasive by a great many people in the scientific and other communities (e.g., economics).

The panel's tentative conclusions on this are (1) the move toward powerful integrated environments is technologically inexorable and unquestionably valuable, (2) the question, then, is how to mitigate the entanglement problems caused by such an approach, and (3) the solution is likely to be incorporating tools to generate and export implementation-independent characterizations of the conceptual model, method of simulation, and depiction of experimental context. The DoD should invest in understanding what is feasible here, what it might request or require, and what incentives would make such things feasible. Best-practices manuals might also prove quite useful, especially those with detailed examples.

Ontologies

Over the last decade, a great deal of research has gone into the development of ontologies. The applications include artificial intelligence, including that for autonomous systems, decision support, and many other examples. Ontology work is likely to prove quite important in the advancement of composability as well, since it is a key element in addressing semantic issues—by standardization in some cases and by agile transformation of representations in others. There is a rich literature on the subject of model ontologies, but we mention here only one example, the Web Ontology Language (OWL), which is under development by the World Wide Web Consortium (www.w3.org/203/Owl).

[5]For expert programmers, even sketches of Java code can be effectively more like pseudocode than something locking people into a particular language.

[6]Recently, suggestions have been made about treating "context" separately and formalizing it as an essential aid to efforts on composability (Yilmaz, 2004). Among the purposes is to increase the odds of recognizing important assumptions affecting sound use of a component. The article, and prior work with Tuncer Oren, also encourages further work on introspective, reflective models—i.e., models that have the ability to report on their own limitations and validity. Some of the most ambitious notions for such work have been discussed under the rubric of "wrapping" (Landauer and Bellman, 1996) in work to make models have surprisingly strong concepts of themselves and the ability to report on that. This goes well beyond what is ordinarily meant by the concept of "wrapper."

Standardization of Representations

Even in the autumn of 2003, it was evident that there were great opportunities for DoD to exploit recent commercial developments that are generating de facto representational standards. The best known of these is the Model Driven Architecture MDA work of the Object Management Group (http://www.omg.org). One recent manifestation of this is the recent release of the second edition of a well-known text for modeling and design (Blaha and Rumbaugh, 2005), which has been substantially rewritten so as to use UML2 rather than the diagrammatic notation of the original edition. Enthusiasm for UML representation is strong and growing; it is a trend that DoD should join, support, and either influence or augment. Many of the issues of simulation composabilty are not solved by UML as it now exists, but—as so often happens—enthusiasms run high and shortcomings are often not mentioned. The committee believes that the concepts embodied in UML methods should be augmented by more detailed specification methods necessary in simulation, such as the DEVS methods developed at the University of Arizona (Ziegler et al., 2000) or the Systems Modeling Language (SysML) being developed by the Object Management Group, and that such augmentation will prove valuable to composability and interoperability. At a different level of detail, a strong base in both computer science and technology now exists for the mechanisms necessary to compose models. Much of this is associated with the Extensible Modeling and Simulation Framework (XMSF).[7]

Retrodocumentation

A continuing challenge for DoD composability is that many (even most) of the relevant existing models are poorly documented legacy code. Although the temptation exists to postulate new models for everything, the reality is that legacy models will be around for a very long time. This suggests to the panel that the DoD should define and invest in a substantial program of retrodocumentation, with or without the cooperation of the original developers, who may have moved on or who may have proprietary concerns.

Conceptual models can be constructed after the fact, and modern representational techniques make it likely that the results would have enduring value if the efforts were done well. Doing the job well could include systematically uncovering deeply buried assumptions about appropriate contexts for the models' use and about phenomena that may not be correctly described. Some of the methods that could be adopted here include these:

- Developing and testing a best-practices guide on how to conduct reviews designed to uncover hidden assumptions.[8]
- Developing pockets of expertise in providing independent consulting and advice on such matters. Analogies exist to current-day teams that provide independent verification and validation, and to "red teams" that conduct independent tests of organizations' information systems.

Improved Wrappers and Metadata

Building on the lessons from the retrodocumentation efforts, among others, it should be possible to develop wrappings for legacy models that will be significantly better than the interfaces currently developed for interoperability purposes and to include in them mechanisms for self-monitoring.[9] For example, models could report when they are being fed inputs that are outside the realm of acceptable domain values or when they see internal state variables taking on values that are either implausible or indicative of operating regimes for which the models are unreliable. Such measures are potentially open-ended, of course, but the committee speculates that much could be accomplished with relatively modest effort. "The best" should not be the enemy of "the much-better-than-now."

Reprogramming

DoD should also be willing, in a few high-leverage cases, to pay for reprogramming legacy models that appear to be substantively valuable but technologically obsolete in troublesome ways. Unfortunately, reprogramming can be quite expensive,[10] so it should not be undertaken lightly. Wrapping methods are often better for near to midterm

[7]See, for example, the Web site of the Naval Postgraduate School's MOVES Institute and related papers (Brutzman et al., 2002).

[8]Uncovering the unstated assumptions of an organization's strategy or plan has become a well-developed methodology. Some of the same methods could be applied. For example, in assumptions-based planning (Dewar, 2003), one uncovers the assumptions and identifies those that are potentially most critical or "load-bearing." Even if they currently appear well based, it is useful to create mechanisms for monitoring the situation so that warnings and adaptations can be made when and if conditions change. In the modeling context, we see the possibility of doing analogous things with metadata and wrappings.

[9]Normal wrappings are merely interfaces that provide orderly and comprehensible mechanisms at the public interface for manipulating what may be complex internal mechanisms of older models or models from which only certain information is to be used in composition. More advanced wrappings, however, can include diagnostics (Landauer and Bellman, 1996, 1999).

[10]It is worth noting, however, that the cost of reprogramming can sometimes be far less than originally estimated, at least if first-team programmers are enlisted. One example of this involved reprogramming the Army's Janus system, which was done quickly and well and made it possible for the Army to have competition and choices of platform as it moved forward.

patches, and substantial redesign is often better if the result is to have enduring value.

Rethinking the Ground Rules and Incentives

One vexing issue that frequently arises in seeking to achieve interoperability and composability is that of proprietary content. Companies that develop models and simulations invest a great deal of time and money in doing so, and the results reflect intellectual capital, which they commonly resist sharing. Programs become proprietary, with the compromise being that the programs have public interfaces necessary for required sharing, as in a federation of models. The remainder of the programs, however, is hidden. This practice, interestingly, extends even to organizations within large companies, where the various divisions may be only modestly more willing to share details with other company divisions than with outsiders.[11] And, of course, the quality of model documentation is notoriously poor.

From a scientific perspective, this situation is appalling—the antithesis of open exchange—but the reasons for it are obvious. Much effort in recent years has gone into finding compromises that call for more of the details to be visible (gray-box approaches, rather than black-box approaches), but complete visibility and the opportunity to make modifications wherever needed is much more rare.

The gray-box approach can accomplish a great deal and is far superior to models being provided as black boxes with only narrow public interfaces. Nonetheless, the panel believes that DoD (and other government agencies) should rethink this entire subject and consider changes of policy and practice that would greatly increase openness. A great deal of empirical evidence exists on which to draw in reassessing. The open-source movement in software, with such triumphs as the UNIX and LINUX systems, is one obvious example, but others exist as well. Models and simulations developed at national laboratories, universities, and not-for-profit organizations have often been much more open.[12] In the business world, moreover, it is not uncommon for an organization that has a simulation developed for it to demand that its full source code and documentation be deliverable, and that it, as the recipient organization, have full ownership and rights. Again, there are many variations, as when a developer makes available good documentation and very extensive mechanisms for modification, but hides—and retains ownership of—underlying machinery that it uses whenever it develops a comparable simulation for a client. This is common practice for financial programs.

This report makes no general recommendations on what DoD should seek to do on this general subject; any effort to impose a one-shoe-fits-all approach might be disastrous. Furthermore, decrees about openness would make no sense unless they were accompanied by fair and appropriate financial incentives and contractually binding legal language. These would require considerable thought based on experience as well as anticipation of behaviors. Nonetheless, the baseline reality is rather odious and is distinctly unhelpful if the DoD wishes to improve composability, reuse, and competition.

Theoretical Research

Based on the foregoing discussions, more theoretical work is needed to understand at least the following:

1. Measures of potential composability that address the time and effort required and that consider the potential need for adaptations (Bartholet et al., 2005)
2. Methods to estimate the reasonable cost of retrodocumentation, development of fewer and more intelligent wrappings, and even reprogramming to create suitable modularity.
3. Methods to improve standardized representation of models and simulations, leaning as heavily as possible on the ongoing industry-sponsored activities but augmenting them as necessary. Currently, the representational methods are to some extent asserted to be good, without a solid description of their strengths and limitations.
4. Achieving sound, mutually informed and calibrated families of models, simulations, and other sources of knowledge (Davis et al., 2005).

Methods for formalizing in practical ways the issues of pragmatics and assumptions and for best assuring that as many such issues as possible are addressed well in documentation and metadata.

REFERENCES

Bartholet, R.G., D.C. Brogran, and P.F. Reynolds. 2005. "The computational complexity of component selection in simulation reuse.'' *Winter Simulation Conference 2005*. Orlando, Fla.

Blaha, Michael, and James Rumbaugh. 2005. *Object-Oriented Modeling and Design with UNL*, 2nd ed. Upper Saddle River, N.J.: Pearson Prentice Hall.

[11]Even when the profit motive is not a consideration, organizations may refuse to share source code for several reasons. One is a desire to maintain tight configuration control so as to maintain quality and standardization. Another is the lack of enthusiasm for revealing imperfections: Computer programs are sometimes clumsy assemblages with less-than-first-rate coding. A third reason is simply that knowledge is power. Even if no direct financial benefit is to be had, there can be substantial indirect benefits from having unique knowledge and expertise.

[12]A starting point for such an assessment might include old standby simulations such as Janus, TACWAR, EADSIM, and more recent systems such as MODSIM. The rules governing distribution of source code, access to it, and ability to make modifications have varied considerably, thereby producing different examples to consider.

Brutzman, Don, Michael Zyda, J. Mark Pullen, and Katherine L. Morse. 2002. "Extensible modeling and simulation framework (Xmsf): Challenges for Web-based modeling and simulation: Findings and recommendations." *Technical Challenges Workshop, Strategic Opportunities Symposium.* Monterey, Calif.

Davis, Paul K., and Robert H. Anderson. 2003. *Improving the Composability of Department of Defense Models and Simulations*. Santa Monica, Calif.: RAND.

Davis, Paul K., Jonathan Kulick, and Michael Egner. 2005. *Implications of Modern Decision Science for Military Decision Support*. Santa Monica, Calif.: RAND.

Dewar, James. 2003. *Assumptions-Based Planning*. London, England: Cambridge University Press.

Hofmann, Marco. 2004. "Challenges of model interoperation in military simulations." *Simulation* 80(12):659-667.

Landauer, Chris, and Kirstie L. Bellman. 1996. "Constructed complex systems: Issues, architectures, and wrappings." *Proceedings EMCSR 96: Thirteenth European Meeting on Cybernetics and Systems Research, Symposium on Complex Systems Analysis and Design*, Vienna, pp. 233-238.

Landauer, Christopher, and Kirstie L. Bellman. 1999. "Lessons learned from wrapping systems." *Fifth International Conference on Engineering of Complex Computer Systems (ICECCS).* Los Alamitos, Calif.: IEEE Computer Society Press, pp. 132-142.

National Research Council (NRC). 1997. *Modeling and Simulation.* Vol. 9 of *Technology for the United States Navy and Marine Corps: 2000-2035.* Washington, D.C.: National Academy Press.

Page, Ernest H., Richard Briggs, and John A. Tufaralo. 2004. "Toward a family of maturity models for the simulation interconnection problem." *Proceedings of the Spring Simulation Operability Workshop.* 045-SIW-145.

Petty, Mikel D., and Eric W. Weisel. 2003. "A composability lexicon." *Proceedings of the Spring Software Interoperability Workshop.* Orlando, Fla.

Szyperski, Clemens. 2002. *Component Software: Beyond Object-Oriented Programming,* 2nd ed. New York, N.Y.: Addison-Wesley.

Turnitsa, Charles. 2005. *Extending the Levels of Conceptual Interoperability Model.* Summer Simulation Conference. Cherry Hill, N.J.

Yilmaz, Levent. 2004. "On the need for contextualized introspective models to improve reuse and composability of defense simulations." *Journal of Defense Modeling and Simulation* 1(3):141-151.

Zeigler, Bernard, Herbert Praenhofer, and Tag Gon Kim. 2000. *Theory of Modeling and Simulation,* 2nd ed. San Diego, Calif.: Academic Press.

D

Committee Membership Biographies

Sallie Keller-McNulty, *Chair*, is dean of Rice University's George R. Brown School of Engineering. She previously headed the Statistical Sciences Group at Los Alamos National Laboratory, where she led a wide range of research and development (R&D) into model validation, reliability, defense analysis, and other topics. Before moving to Los Alamos, Dr. Keller-McNulty was professor and director of graduate studies at the Department of Statistics, Kansas State University, where she had been on the faculty since 1985. She spent 1994-1996 as a program officer in the National Science Foundation's Division of Mathematical Sciences. Her ongoing areas of research focus on computational and graphical statistics applied to statistical databases, including complex data/model integration and related software and modeling techniques, and she is an expert in the area of data access. She has served on the Information Technology panel of the National Institute of Standards and Technology Board (2001-2004); the Committee on National Statistics' Panel on Research for Future Census Methods (1999-2004); the Board on Mathematical Sciences and their Applications (2000-2003); the Committee on Applied and Theoretical Statistics (Chair, 2000-2003); and the Computer Science and Telecommunications Board's Committee on Computing and Communications Research to Enable Better Use of Information Technology in Government (1998-2002). She is a national associate of the National Academy of Sciences and fellow of American Association for the Advancement of Science (AAAS). She received her Ph.D. in statistics from Iowa State University of Science and Technology. She is a fellow of the American Statistical Association (ASA) and has held several positions within the ASA, including, currently, as president. She is an associate editor of *Statistical Science* and has served as associate editor of the *Journal of Computational and Graphical Statistics* and the *Journal of the American Statistical Association*. She served on the Executive Committee of the National Institute of Statistical Sciences, on the Executive Committee of AAAS Section U, and chairs the Committee of Presidents of Statistical Societies, of which she is a former president.

Kirstie L. Bellman recently returned to the Aerospace Corporation after 4 years at the Defense Advanced Research Projects Agency (DARPA) to start a new bicoastal R&D center called the Aerospace Integration Sciences Center (AISC). The center serves as R&D capability for a number of the Department of Defense (DoD) and government agencies. AISC's focus is on the development of advanced system and model integration methods, new analytic techniques, and evaluation tools for assessing the impacts of new technologies. Upon completion of her term at DARPA, Dr. Bellman received an award from the Office of the Secretary of Defense for excellence in her programs. Dr. Bellman has over 35 years of academic, industrial, and consulting experience in both laboratory research and the development of models and information architectures for large military and government programs. Her published research spans a wide range of topics in cognitive science, neuroscience, and computer science. Her recent work focuses on the use of domain-specific languages and formally based architectural description languages to design and analyze information architectures and reflective architectures that use models of themselves to manage their own resources and to reason about appropriate behavior. In this work, she has also been developing methods for modeling the organizational and technical aspects of complex systems. With a number of academic partners, she is also developing new mathematical approaches to the analysis of virtual worlds that contain collaborating humans, artificial agents, and heterogeneous representations, models, and processing tools. Dr. Bellman is an elected fellow of the AAAS.

Kathleen M. Carley is professor of computation, organizations, and society at the Institute for Software Research in the School of Computer Science at Carnegie Mellon University. She also heads Carnegie Mellon's Center for Computational Analysis of Social and Organizational Systems. Dr. Carley received a Ph.D. in sociology from Harvard University. Her research combines cognitive science, social networks, and computer science. Her specific research areas are computational social and organization theory; group, organizational, and social adaptation and evolution; dynamic network analysis; computational text analysis; and the impact of telecommunication technologies and policy on communication, information diffusion, disease contagion, and response within and among groups particularly in disaster or crisis situations. Her models meld multiagent technology with network dynamics and empirical data. She has developed a number of tools for extracting (AutoMap), analyzing (ORA), and reasoning about change in (Construct, DyNet) social and knowledge networks. Three of the large-scale multiagent network models she and her group have developed are BioWar, a city-scale model of weaponized biological attacks; OrgAhead, a model of strategic and natural organizational adaptation; and Construct, a model of the coevolution of social and knowledge networks and personal/organizational identity and capability. She is the founding coeditor of the journal *Computational and Mathematical Organization Theory* and has coedited several books in the multiagent and dynamic social networks area. She was a member of the National Research Council's (NRC's) Panel on Modeling Human Behavior and Command Decision Making: Representations for Military Simulations.

Paul K. Davis is a senior scientist and research leader at RAND and a professor of policy analysis in the RAND Graduate School. His current research relates to strategic planning; high-level decision support; representing adversary reasoning; capabilities-based planning; effects-based operations; deterrence in the counterterrorism era; military transformation; advanced methods for modeling and simulation, including model composability; and missile defense. Dr. Davis teaches graduate courses in defense planning, counterterrorism policy, and policy analysis of strategy problems with massive uncertainty. Dr. Davis has served on a number of studies for the Defense Science Board. He was awarded the Vance R. Wanner award by the Military Operations Research Society for lifetime achievement. Before joining RAND, Dr. Davis was a senior executive in the Office of the Secretary of Defense. He holds a B.S. from the University of Michigan and a Ph.D. in chemical physics from MIT. He was at one time a member of the NRC's Naval Studies Board and served on several of its committees.

Richard Ivanetich is Institute Fellow at the Institute for Defense Analyses (IDA), having been appointed to that position in 2003. His experience spans a number of areas of defense systems, technology, and operations analyses, relating primarily to computer and information systems, command-and-control systems and procedures, modeling and simulation of systems and forces, crisis management, and strategic and theater nuclear forces. His previous positions at IDA include director of the Computer and Software Engineering Division (1990-2002) and assistant director of the System Evaluation Division (1985-1990). Prior to joining IDA in 1975, Dr. Ivanetich was assistant professor of physics at Harvard University (1969-1974). He has served on numerous scientific boards and advisory committees, among them the NRC's Naval Studies Board (1998-2004) and the DARPA Information Science and Technology study group (1990-2004).

Kathryn B. Laskey is a professor of systems engineering and operations research at George Mason University. She received her master's degree in mathematics from the University of Michigan and her Ph.D. in statistics and public affairs from Carnegie Mellon University. Dr. Laskey studies Bayesian inference and decision theory, multisource fusion, uncertainty in artificial intelligence, and situation assessment. Her broad research interest is the use of information technology to support better inference and decision making. Within this area, her interests lie in understanding the proper role of normative, behavioral, and computational theories in the modeling and support of decision making. Dr. Laskey is a research fellow with the Krasnow Institute for Cognitive Science at George Mason University (GMU) and associate director of the Center of Excellence in Command, Control, Communications and Intelligence at GMU. She has been on several NRC committees, including the committee that wrote the 2001-2002 study on scientific evidence on the polygraph.

R. Bowen Loftin is vice president and chief executive officer of Texas A&M University at Galveston. He has worked extensively in the academic community as a researcher, instructor, and mentor. In 1977 he joined the faculty of the University of Houston-Downtown. In 1994 he founded the NASA/University of Houston Virtual Environments Research Institute and in 1999 became the chair of the Department of Computer Science at the University of Houston. He received the University of Houston-Downtown Excellence in Teaching Award in 1982 and the Excellence in Service Award in 1984 and 1985. For the past 5 years he has served as executive director of a research center at Old Dominion University in Virginia while also managing the university's multidisciplinary graduate degree programs in modeling and simulation. Under his tenure there, the master's student enrollment increased to over 50 students and the doctoral student population increased from 2 to over 40 students. His major field of research is modeling, analysis, and simulation. He holds Ph.D. and M.A. degrees in physics from Rice University and a B.S. in physics from Texas A&M University. He has written numerous articles on topics as varied as artificial intelligence, virtual reality, and computer-aided training and simulations. He has won multiple awards, including various NASA awards for inventions, publications, and public service in 1989, 1992, 1993, 1994, and 1995.

GEN David M. Maddox (retired), U.S. Army, is a retired four-star general who currently works as a consultant. A member of the National Academy of Engineering (NAE), GEN Maddox has expertise in operations research, simulation and modeling, logistics, joint operations/warfighting, organizational design, and materiel requirements. He retired from the U.S. Army in 1995 after serving as commander in chief, U.S. Army in Europe. Since that time, he has performed extensive consulting work on concepts, systems requirements, analytic techniques and analyses, operations and systems effectiveness, and program capture strategies to civilian corporations, government agencies, and defense industries. GEN Maddox has had extensive command experience. He served four tours in Germany, during which he commanded at every level from platoon through Army group and theater. Following command at platoon and troop level in the 14th Armored Cavalry Regiment (ACR), he commanded the 1st Squadron, the 11th ACR in Fulda, the 2nd ACR (he was the 61st Colonel of the Regiment) in Nuremberg, the 18th Infantry Division (mechanized) in Bad Kreuznach, the V Corps in Frankfurt, and NATO's Central Army Group and the U.S. Army in Europe, Heidelberg, and the 7th Army in Heidelberg.

Dennis K. McBride is president of the nonprofit Potomac Institute for Policy Studies. He was a career Navy scientist specializing in modeling and simulation, particularly cognitive modeling, maintenance ergonomics modeling, and tactical air system discrete event modeling. He served as a DARPA program manager and a program officer at the Office of Naval Research. Dr. McBride helped to write the legislation that created the Defense Modeling and Simulation Office, the sponsor of this study. He chaired the NRC Panel on Engineering for Complex Systems (2002-2003) and was a member of the NRC committee that reviewed NASA's Pioneering Revolutionary Technology (PRT) Program (2002-2003).

COL Michael McGinnis joined Old Dominion University in June 2006 as the executive director of the Virginia Modeling, Analysis and Simulation Center. Prior to assuming this position Brigadier McGinnis served for 7 years as professor and head of the Systems Engineering Department, U.S. Military Academy at West Point. His previous Army modeling, simulation, and analysis assignments include director of the U.S. Army Unit Manning Task Force, director of the U.S. Army TRADOC Analysis Center at the Naval Postgraduate School in Monterey, California, and director of the U.S. Military Academy Operations Research Center. He has served on key government engineering, modeling, simulation, and analysis committees to bring about change at the Army and DoD levels. Dr. McGinnis is a graduate of the U.S. Military Academy and has M.S. degrees in applied mathematics and operations research from Rensselaer Polytechnic Institute and a Ph.D. from the University of Arizona in systems and industrial engineering. He attended the Command and General Staff College at Fort Leavenworth and the Naval War College in Newport, Rhode Island, where he earned an M.A. in national security and strategic studies. Dr. McGinnis's professional and scholarly body of work includes three national awards and over 40 published and peer-reviewed papers published during 17 years of working in the fields of systems engineering and operations research. Dr. McGinnis has been honored with the 1995 Military Operations Research Society Rist Prize, the 2004 Military Operations Research Society Barchi Prize, and the best paper award for the 2005 Interservice/Industry Training, Simulation and Education Conference Research and Development Track.

Stephen Pollock is Herrick Emeritus Professor of Manufacturing and Emeritus Professor of Industrial and Operations Engineering at the University of Michigan. He taught courses in decision analysis, mathematical modeling, dynamic programming, and stochastic processes. Dr. Pollock's recent research has included developing cost-optimal monitoring and maintenance policies, sequential hypothesis testing, modeling large multiserver systems, and adaptive optimization of radiation treatment plans under uncertainty. He is the recent past director of the University of Michigan's Program in Financial Engineering and its Engineering Global Leadership honors program. He has served as area editor of *Operations Research*, senior editor of *IIE Transactions*, president (1986) of the Operations Research Society of America, and a senior fellow of the University of Michigan Society of Fellows. Dr. Pollock is a founding fellow of the Institute for Operations Research and the Management Sciences, was awarded its Kimball Medal in 2002, and was elected to the NAE in 2003. Among his many NRC activities, he chaired a recent study on the test and evaluation plans for the Army's Stryker family of vehicles.

David R. Pratt is the chief scientist (fellow) for Science Applications International Corporation's (SAIC's) Strategies Simulation and Training business unit. As a vice president for technology, his responsibilities include developing and fostering continued leading-edge information technology and modeling and simulation technologies. He provides both strategic and tactical guidance in technical and programmatic matters. With a research base of over $6 million per year, he oversees both internal and external research projects. Recently, these research projects have included robotics, evolutionary algorithms, synthetic agent behaviors, language performance studies, data management and distribution, data warehousing and mining, user interface, and multithreading/multiprocessing. Dr. Pratt also serves as the forces modeling and simulation point of contact for DoD's High Performance Computing Modernization Program. Before joining SAIC, Dr. Pratt was the technical director for the largest simulation software effort ever undertaken by the DoD, the Joint

Simulation System. Formerly a tenured associate professor of computer science at the Naval Postgraduate School and adjunct teaching instructor at the University of Central Florida, he has an extensive academic background that includes over 50 publications and $5 million of external academic research funding.

Stephen M. Robinson is professor of industrial and systems engineering at the University of Wisconsin-Madison, where he has been a member of the faculty since 1972. He has a collateral appointment as professor of computer sciences, and has held administrative appointments as chair of the Department of Industrial Engineering and as assistant director of the Mathematics Research Center. His research specialty is mathematical programming (methods for making the best use of limited resources, applied in logistics, transportation, manufacturing, and many other areas). He is author, coauthor, or editor of seven books and 91 scientific research papers and has directed numerous funded research projects at the University. His research accomplishments have been recognized by the award of the honorary doctor's degree from the University of Zürich, Switzerland, the George B. Dantzig Prize of the Mathematical Programming Society and the Society for Industrial and Applied Mathematics, and the John K. Walker, Jr., Award of the Military Operations Research Society. He is a fellow of the Institute for Operations Research and the Management Sciences (INFORMS). Dr. Robinson has been an elected member of the councils of the Operations Research Society of America (now INFORMS) and of the Mathematical Programming Society, and he also served for 4 years as secretary and, concurrently, as a member of the board of directors of INFORMS. He has also been an editor of several scientific journals and has served on numerous governmental and professional advisory committees. He is a former trustee of the Village of Shorewood Hills, Wisconsin, and from 1991 to 2002 he served on the board of overseers of Simon's Rock College, Great Barrington, Massachusetts. Dr. Robinson is also a retired colonel in the Army of the United States and a graduate of the U.S. Army War College. Prior to joining the University of Wisconsin-Madison he served on active duty for 6 years as a regular Army officer. He is a current member of the NRC's Board on Mathematical Sciences and Their Applications.

Detlof von Winterfeldt is a professor of public policy and management in the School of Policy, Planning, and Development (SPPD) at the University of Southern California (USC) and the director of USC's Center for Risk and Economic Analysis of Terrorist Events (CREATE). For the past 25 years, he has been active in teaching, research, university administration, and consulting. He has taught courses in statistics, decision analysis, risk management, and human judgment and decision making. His research interests are in the foundation and practice of decision and risk analysis as applied to technology, the environment, and national security problems. He is the coauthor of two books and author or coauthor of over 100 articles and reports on these topics. His administrative experiences include serving as deputy dean of SPPD, as director of USC's Institute for Civic Enterprise, and as chairman of USC's Systems Science Department. As a consultant he has applied decision and risk analysis to many management problems of government and private industry. In 2000, he received the Ramsey Medal for distinguished contributions to decision analysis from the Decision Analysis Society of INFORMS. He is a fellow of INFORMS and of the Society for Risk Analysis. Dr. von Winterfeldt received his Ph.D. from the University of Michigan, Ann Arbor, in mathematical psychology. He has served on a number of NRC study committees and recently finished a term on the Board on Mathematical Sciences and Their Applications.

Michael Zyda is the director of the GamePipe Laboratory at USC's Viterbi School of Engineering, a professor of engineering practice in the USC Department of Computer Science, and a staff member of USC's Information Sciences Institute, located in Marina del Rey, California. From fall 2000 to fall 2004, he was the founding director of the MOVES Institute, located at the Naval Postgraduate School (NPS) in Monterey, and a professor in the Department of Computer Science at NPS as well. From 1986 until the founding of the MOVES Institute, he was the director of the NPSNET Research Group. Dr. Zyda's research interests include computer graphics; large-scale, networked 3-D virtual environments; agent-based simulation; modeling human and organizational behavior; interactive computer-generated story, modeling and simulation; and interactive games. He is a pioneer in computer graphics, networked virtual environments, modeling and simulation, and serious games. He is a member of the Academy of Interactive Arts and Sciences. He served as the principal investigator and development director of the America's Army PC game funded by the assistant secretary of the army for manpower and reserve affairs. He took America's Army from conception to three million plus registered players, transforming Army recruiting. Dr. Zyda chaired a major NRC study that examined the potential interface between the entertainment industry and the military.

E

Acronyms

ACT	Allied Command Transformation
ACTD	Advanced Concept Technology Demonstration
BCNPS	BattleSpace Communications Network Planner and Simulator
CAS	complex adaptive system
CASTFOREM	Combined Arms and Support Task Force Evaluation
CBP	capabilities-based planning
COA	course of action
COTS	commercial off-the-shelf
DARPA	Defense Advanced Research Projects Agency
DDDAS	Dynamic Data-Driven Application System
DEVS	discrete event specification
DIME	diplomatic, intelligence, military, and economic
DMSO	Defense Modeling and Simulation Office, now called the Modeling and Simulation Coordination Office (M&SCO)
DNA	Defense Nuclear Agency; dynamic network analysis
DoD	Department of Defense
EADSIM	Extended Air Defense Simulation
EBO	effects-based operations
EBP	effects-based planning
EMAST	End-to-End Modeling and Simulation Testbed
FAR	flexible, adaptive, and robust
FCRT	Future Capabilities, Research, and Technology
FP	fictitious play
GAO	General Accounting Office
HLA	High-Level Architecture
ICCRTS	International Command and Control Research and Technology Symposium
IEEE	Institute of Electrical and Electronics Engineers
ISAAC	Irreducible Semi-Autonomous Adaptive Combat Model
ITAR	International Traffic in Arms Regulations
JICM	Joint Integrated Contingency Model
JSAF	Joint Semi-Automated Forces
JSIMS	Joint Simulation System
JTF	Joint Task Force
JTIDS	Joint Tactical Information Distribution System
JTRC	Joint Total Relevant Cost
JWARS	Joint Warfare System
M&S	modeling and simulation
M&SCO	Modeling and Simulation Coordination Office (formerly DMSO)

MDA	Model-Driven Architecture
MIT	Massachusetts Institute of Technology
MMOG	massively multiplayer online game
MORS	Military Operations Research Society
MOVES	Modeling, Virtual Environments, and Simulation
MPARS	Multifunction Phased Array System
MRM	multiresolution model
MS&A	modeling, simulation, and analysis
MSIAC	Modeling and Simulation Information Analysis Center
MSRR	Modeling and Simulation Resource Repository
NAS	National Academy of Sciences
NATO	North Atlantic Treaty Organization
NCO	network-centric operations
NCW	network-centric warfare
NETWARS	network warfare simulation
NRC	National Research Council
NSF	National Science Foundation
NSS	Navy Simulation System
OEF	Operation Enduring Freedom
OFT	Office of Force Transformation
OIF	Operation Iraqi Freedom
PCAS	preconflict anticipation shaping
PMESII	political, military, economic, social, infrastructure, and information
PSYOPS	psychological operations
QDR	Quadrennial Defense Review
R&D	research and development
R&S	reconstruction and stabilization
RAID	real-time adversarial intelligence and decision making
RDT&E	research, development, technology, and engineering
ROI	return on investment
RSAS	RAND Strategy Assessment System
SBIR	Small Business Innovative Research
SEU	subjected expected utility
SNA	social network analysis
SOA	service-oriented architecture
SoS	system of systems
STTR	Small Business Technology Transfer
SysML	Systems Modeling Language
TACWAR	Tactical Warfare Model
TENA	Test and Training Enabling Architecture
UARC	university-affiliated research center
UAV	unmanned air vehicle
UGV	unmanned ground vehicle
UML	Unified Modeling Language
USAF	United States Air Force
USC ICT	University of Southern California, Institute for Creative Technologies
UxV	unmanned (air or ground) vehicle
VAST	virtual at-sea training
VIC	vector in commander
VV&A	verification, validation, and accreditation
WOL	Web Ontology Language
XMSF	Extensible Modeling and Simulation Framework